Pocket Book of Technical Writing for Engineers and Scientists

Mcgraw-Hill continues to bring you the *BEST* (**B**asic **E**ngineering **S**eries and **T**ools) approach to introductory engineering education:

Burghardt, *Introduction to Engineering Design and Problem Solving* 0070121885 (GOP)

Chapman, *Fortran 90/95 for Scientists and Engineers,* 2/e 0072922389

Donaldson, *The Engineering Student Survival Guide,* 2/e 0072436336

Eide/Jenison/Northup, *Introduction to Engineering Design and Problem Solving,* 2/e 0072402210

Eisenberg, *A Beginner's Guide to Technical Communication* 0070920451

Finkelstein, *Pocket Book of Technical Writing for Engineers and Scientists* 0072976837

Gottfried, *Spreadsheet Tools for Engineers using Excel* 0072480688

Palm, *Introduction to MatLab 6 for Engineers with 6.5 Update and Additional Topics in Animation, Graphics, and Simulink®* 0072970553

Pritchard, *Mathcad: A Tool for Engineering Problem Solving* 0070121893

Schinzinger/Martin, *Introduction to Engineering Ethics* 0072339594

Smith, *Teamwork and Project Management,* 2/e 0072922303

Tan/D'Orazio, *C Programming for Engineering and Computer Science* 0079136788

Additional Titles of Interest:

Andersen, *Just Enough Unix,* 4/e 0072463775

Eide/Jenison/Mashaw/Northup, *Engineering Fundamentals and Problem Solving,* 4/e 0072430273

Holtzapple/Reece, *Foundations of Engineering,* 2/e 0072480823

Pocket Book of Technical Writing for Engineers and Scientists

Second Edition

Leo Finkelstein, Jr.
Wright State University

Boston   Burr Ridge, IL   Dubuque, IA   Madison, WI   New York
San Francisco   St. Louis   Bangkok   Bogotá   Caracas   Kuala Lumpur
Lisbon   London   Madrid   Mexico City   Milan   Montreal   New Delhi
Santiago   Seoul   Singapore   Sydney   Taipei   Toronto

The McGraw·Hill Companies

Higher Education

POCKET BOOK OF TECHNICAL WRITING FOR ENGINEERS AND SCIENTISTS
SECOND EDITION

2 3 4 5 6 7 8 9 0 DOC/DOC 0 9 8 7 6 5

ISBN 0–07–246849–1

Publisher: *Elizabeth A. Jones*
Senior sponsoring editor: *Carlise Paulson*
Developmental editor: *Kate Scheinman*
Senior project manager: *Sheila M. Frank*
Senior production supervisor: *Sherry L. Kane*
Lead media project manager: *Audrey A. Reiter*
Senior media technology producer: *Eric A. Weber*
Senior coordinator of freelance design: *Michelle D. Whitaker*
Cover designer: *Kelly Fassbinder / Imagine Design Studio*
Compositor: *Lachina Publishing Services*
Typeface: *10 / 12 Century Schoolbook*
Printer: *R. R. Donnelley Crawfordsville, IN*

Cover images: Celltower, computer monitor, cpu fans, and x-ray spectrometer image provided by Leo Finkelstein, Jr.; Chemical Compound Abstract: Photodisc BS35, Techno Abstracts; Keyboard: Photodisc DT5, Keyboards and Computers; Gears: Photodisc DT1, Gears Metaphors of Interaction; Compass/Ruler: Photodisc SS18, Technology Perspectives

Library of Congress Cataloging-in-Publication Data

Finkelstein, Leo, 1946–
    Pocket book of technical writing for engineers and scientists / Leo Finkelstein, Jr. — 2nd ed.
        p.   cm.
    ISBN 0–07–246849–1 (hard copy : alk. paper)
    1. Technical writing—Handbooks, manuals, etc. I. Title.

T11.F53   2005
808'.0666—dc22   2003025059
                                    CIP

www.mhhe.com

This book is dedicated to the memory of my father, who taught me all the important things I needed to know.

About the Author

Leo Finkelstein, Jr., received a bachelor's degree from the University of North Carolina at Chapel Hill in 1968; a master's from the University of Tennessee at Knoxville in 1969; and a Ph.D. from Rensselaer Polytechnic Institute at Troy, New York, in 1978. He is currently Lecturer and Director of Technical Communication for the College of Engineering and Computer Science, Wright State University, Dayton, Ohio. As an associate professor, he directed the technical writing program at the U.S. Air Force Academy while also serving as an adjunct associate professor for the University of Colorado at Colorado Springs. He wrote, produced, and directed technical films in Southern California and commanded a combat-documentation, photographic unit in Southeast Asia during the Vietnam War. In addition, his military service includes experience in both space and logistics systems. He holds FCC commercial and amateur radio licenses, has a black belt in Taekwondo, and is an avid user of all types of gadgets.

Contents

1 Introduction   **1**
What Is Technical Writing?   1
Abstraction   2
Audience and Purpose   5
Report Writing and Audience   7
Notes   10

2 Ethical Considerations   **11**
What Are Ethics in Technical Writing?   12
Plagiarism   17
Image Alteration and Ethics   20
Exercises   21
    Plagiarism   21
    Image Compositing   21
Notes   23

3 Technical Definition   **25**
What Is a Technical Definition?   25
Classifications and Classes   27
Differentiation   29
Avoiding Mistakes   29
Extensions   30
    Further Definition   31
    Comparison and Contrast   31
    Classification   31
    Cause and Effect   31

Process   32
Exemplification   32
Etymology   32
Required Imprecision   33
A Word about Defining Specifications and
Standards   34
Definition Checklist   35
Exercise   36
Notes   37

4 Description of a Mechanism   **39**
What Is a Mechanism Description?   39
Example of a Mechanism Description   40
  Introduction   41
  Discussion   44
  Conclusion   47
Visuals and Mechanism Descriptions   48
Putting It All Together   49
Specifications and Functional Mechanism
Descriptions   53
Threaded Example   54
Mechanism Description Checklist   55
Exercise   56

5 Description of a Process   **59**
What Is a Process Description?   59
Outline of a Process Description   60
Description of a Mechanism in Operation   62
  Introduction   62
  Discussion   63
  Conclusion   66
Visuals and Process Descriptions   66
Putting It All Together   67
Description of a Conceptual Process   70
Putting It All Together   71
Threaded Example   75

Process Description Checklist   77
Exercise   78

6 Proposals   **83**
What Is a Proposal?   83
Formal and Informal Proposals   84
  Formal Proposals   84
  Informal Proposals   87
Layout and Presentation   95
Cover Letters and Title Pages   96
Attachments and Appendixes   97
Putting It All Together   98
Threaded Example   105
Proposal Checklist   108

7 Progress Reports   **111**
What Is a Progress Report?   112
Progress Report Formats   113
  Introduction   113
  Status   116
  Conclusion   117
Putting It All Together   118
Threaded Example   122
Progress Report Checklist   124

**8 Feasibility and Recommendation
  Reports**   **125**
How Do Feasibility and Recommendation
Reports Differ?   125
Writing Feasibility and Recommendation
Reports   127
  Introduction   129
  Discussion   132
  Conclusion   135
Putting It All Together   136
Threaded Example   141

Feasibility and Recommendation Report
Checklist   144

9 Laboratory and Project Reports   145
What Are Laboratory and Project
Reports?   146
Threaded Example   149
Putting It All Together   153
Student Project Report   160
Laboratory and Project Report Checklist   164

10 Instructions and Manuals   167
What Are Instructions?   167
   Instructions for the Layperson   169
   Introduction   169
   Discussion   170
   Conclusion   176
Putting It All Together   176
Threaded Example   185
More on Manuals   188
Instructions Checklist   189
Note   190

11 Research Reports   191
What Are Research Reports?   191
Developing a Research Report   193
   Introduction   193
   Background   195
   Discussion   196
   Conclusion   198
   References and Appendix   198
Putting It All Together   199
Threaded Example   205
Research Report Checklist   210

12 Abstracts and Summaries   211
What Are Descriptive Abstracts?   211
   Writing Descriptive Abstracts   211

What Are Informative Abstracts?   212
 Writing Informative Abstracts   213
What Are Executive Summaries?   214
 Writing Executive Summaries   215

13 Grammar and Style   **229**
Grammar: What Is It and Why Is It
a Big Deal?   229
 Comma Splices   230
 Fused Sentences   231
 Sentence Fragments   232
 Misplaced-Modifier Errors   233
 Passive Voice Problems   234
 Verb Agreement Errors   236
 Pronoun Agreement Errors   237
 Pronoun Reference Errors   238
 Case Errors   238
 Spelling Errors   239
 A Word about Homonyms   240
 Spelling and Numbers   241
 Noun Clauses   242
Stylistic Considerations   243
 Economy   244
 Precision   245

14 Documentation   **247**
What Is Documentation?   248
Documentation Styles   249
When to Document Sources   249
 To Meet Legal Requirements   249
 To Meet Academic Standards   250
 To Establish Credibility   250
How to Document Sources   250
Print Media Examples   253
 Books   253
 Journals   254
 Conference Papers   254
 Encyclopedias   254

Newspapers   254
Nonjournal Entries   255
Technical Reports   255
Dissertations and Theses   255
Electronic Media Examples   255
   Website   257
   Online Forum   258
   FTP Site   258
   Computer Local Storage Media (Computer Disk,
   Flash Card, etc.)   258
Other Examples   258
   Interview   258
   Lecture   259
Checklist for Documentation   259
Notes   259

15 Visuals   **261**
What Are Visuals?   261
Guidelines for Design of Visuals   262
   Reproducibility   262
   Simplicity   263
   Accuracy   263
Types of Visuals   263
   Diagrams   263
   Graphs   264
   Schematics   269
   Tables   270
   Photographs   273
   Image Alteration   275
   Process Limitations   276
Conclusion   277
Checklist for Visuals   277

16 Electronic Publishing   **279**
What Is Electronic Publishing?   280
Common Electronic Publishing
File Formats   280
   Document Files   280

Graphics Files   282
Hyperlinked Documents   283
Producing Hypertext Documents   285
Converting Traditional Documents
to Hypertext   285
How Web Browsers Work   287
Guidelines for Organizing Hypertext
Documents   288
About Electronic Publishing and
Copyright   293
Checklist for Electronic Publishing   294
Notes   295

17 Presentations and Briefings   297
What Are Presentations and Briefings?   297
Substantive Ideas   298
Clear, Coherent Organization   298
Terminology and Concepts   299
Effective Delivery   300
Speaking Situations   301
Impromptu   301
Extemporaneous   302
Manuscript   303
Speaking Purposes   303
Informative   304
Demonstrative   304
Persuasive   304
Technical Briefings   305
General Guidelines   305
Title Chart   307
Overview Chart   308
Discussion Charts   309
Summary Chart   309
Concluding Chart   310
Controlling Complexity   311
Visuals and Complexity   311
Special Effects   312
Checklist for Presentations   313

18 Resumes and Interviews   315

What Is a Resume?   316

Writing a Resume   317

   Objective   318

   Strengths   319

   Education   320

   Computer Skills   321

   Experience   322

   Personal   323

Ten Tips for Creating a Good Resume   323

Cover Letters   325

Finding Jobs on the Internet   328

Putting It All Together   328

Interviewing   331

Cover Letter Checklist   334

Resume Checklist   334

Interview Checklist   334

Notes   335

19 Team Writing   337

Student versus Professional Team
Writing   338

The Process of Team Writing   339

   Requirements   340

   Preliminary Actions   342

   Document Production   343

Example of Professional Team Writing   345

   Requirements   346

   Preliminary Actions   347

   Document Production   348

Student Team Writing   349

   Requirements   350

   Preliminary Actions   351

   Document Production   352

Conclusion   353

Index   355

Preface

I did not envision the first edition of *Pocket Book of Technical Writing for Engineers and Scientists* as being your typical technical writing textbook—and the same holds true for the second edition. Although this new edition provides more information than its predecessor, it still avoids the sterile, encyclopedic treatment of writing concepts that few students want or need.

    So why did I write a textbook like this? Over the years, I've found that engineering and science students who are required to buy technical writing texts rarely use them—at least not effectively. That's so because many traditional texts are written for professors, not students—and are long on theory and short on practical information. I believe these books too often provide a nonproductive, frustrating experience for many who are just trying to find effective guidance to satisfy their most pressing technical writing requirements.

Purpose

My goal is to provide a practical, down-to-earth, technical writing guide for busy students with packed curricula and demanding requirements. You'll find that this book doesn't take itself too seriously, and it doesn't spend "forever and a day" getting to the point. You'll also find that this book

Approach

is oriented toward readers who not only study but also enjoy science and technology.

This book provides engineering and science students with straightforward, practical solutions that will be easy and painless to use for meeting a wide range of technical writing challenges, whether in the classroom or the workplace. The emphasis throughout is on sensible, real-world problem solving. If you're looking for esoteric discussions of linguistic relativity or transformational grammars, look somewhere else! You won't find that kind of material in this book. What you will find is the information that you can use, presented in a way that you can use it.

Organization

I organized the second edition around three major sections: the first one to discuss basic material, the second to describe how to write the most common technical documents, and the third to provide useful information that doesn't fit in the first two sections. Here's a brief, section-by-section breakdown.

Section I: Technical Writing Basics

Chapters 1 through 5 deal with basic considerations, including the component skills you'll need to produce effective technical writing. Expanding on the approach used in the first edition, this section includes

- Guidance on how to define terms and describe mechanisms and processes, because virtually any kind of technical writing requires definitions and descriptions. If you can explain to someone what something is, what it looks like, and how it works, you have the basics for being an effective technical writer.

- An expanded, practical discussion of ethics as it relates to technical writing. I have included not only a general framework from which to view ethics, but also an expanded discussion of the two serious problem areas in technical writing today: *plagiarism* and *image alteration*.

Section II: Technical Documents

Chapters 6 through 12 provide a practical guide for producing the most common technical documents. The second edition introduces several innovations while retaining the basic approach of its predecessor. As in the first edition, I have incorporated fictitious, often humorous, theoretically correct examples; and I have developed these examples step by step to show you how to write the most common technical documents. Substantial new material has also been added to update and enhance this approach, including

- Discussions of the most common technical documents, including proposals, progress and status reports, feasibility and recommendation reports, laboratory and project reports, instructions and manuals, research and state-of-the-art reports, and abstracts and summaries. The examples have been reworked to touch on a broader range of disciplines and interests.
- A clearer differentiation between expert and general technical audiences, often with separate examples. Where useful, I also differentiate between professional technical writing and student technical writing.
- An additional Threaded Example that uses a single piece of fictitious, but theoretically correct, technology throughout the book as the core subject matter for various technical writing documents and techniques.

Section III: Other Useful Stuff

Chapters 13 through 19 provide additional material you'll need when writing technical documents—from enhancing your grammar and style, to producing presentations and briefings. This section includes

- An updated, expanded treatment of how to write properly in a style that will get your point across effectively and economically.
- Clear instructions for how to organize and produce presentations and briefings that will get the job done.
- A simple but effective approach for providing source documentation that doesn't require hours with a stuffy style manual.
- A straightforward treatment of visuals in technical writing. Clear discussions are provided for when and how to use charts, graphs, diagrams, tables, and photographs, including the proper use of photograph alteration.
- A practical, effective approach for producing resumes and cover letters, along with several tips for going on job interviews.
- A real-world, practical, extended treatment of team writing in both professional and student environments.

Conclusion   I believe I have approached this book in a way that strikes the right balance for modern engineering and science students between the amount of information provided and the practical usability of the book. I wanted enough information to make the book useful, but not so much as to make it hard to use. With the outlined models provided in many of the chapters, you will be able to ascertain quickly what is required for the specific kind of writing you're doing. Reading through the examples as they are developed will give you further insight into the process of actually producing a technical document. Looking at how the document appears after it has been put together should further clarify basic layout considerations, along with how to properly integrate supporting visuals with your text discussions. Finally, using the checklists will do for you what it does for airplane pilots: It will help you avoid the kinds of natural, human errors that can be costly or even catastrophic.

Being an effective technical writer is becoming increasingly important, especially in our modern, high-technology society. If you cannot communicate what you know to those who need to know it, then what you know will not count for much. In engineering and science, being able to put your ideas into a form that others can use easily and effectively is the key to success—yours and theirs. In that regard, I sincerely hope you can put the second edition to good use.

Acknowledgments

Acknowledging those who helped me with both the first and second editions is no simple task, because so many tried so hard to make me, or at least my book, look good. First and foremost, I thank my wife and best friend, Phyllis A. Finkelstein, whose nonstop encouragement and superb proofreading were nothing short of miraculous. And yes, many of her friends still say that her being married to me for more than 34 years is also nothing short of miraculous.

I thank my friend and colleague, William G. Dwyer, a senior analyst, retired Air Force colonel, current technical writing teacher, and onetime Shakespeare scholar, for his encouragement, insight, and patient proofreading of my writing. I thank my friend and colleague, Thomas A. Sudkamp, professor of computer science and engineering, for plowing through the entire manuscript and rooting out so many embarrassing errors in logic and style. I thank my friend and colleague, Fred D. Garber, associate professor and chair of electrical engineering, for his patience in listening to my crazy ideas and pointing me in creative directions that were theoretically sound.

I thank my son, Stephen B. Finkelstein, a professional computer scientist, for "telling Dad the truth" when no one else would; and I thank my former student and longtime friend, Sharon E.

Liebel, a biomedical engineer, ergonomics expert, and respected manager, for providing, as always, her unique perspective and insight.

I also express my gratitude to the following reviewers for their most helpful comments: Kenneth J. Breeding, The Ohio State University; John Brocato, Mississippi State University; Barton B. Cregger, Virginia Commonwealth University; Thomas Grimm, Michigan Technological University; Thimios Jordanides, California State University, Long Beach; Suzanne Karberg, Purdue University; Steven Kunert, Oregon State University; Jeanne Linsdell, San Jose State University; Scott J. Mason, University of Arkansas; Arun Nevader, University of California, Berkeley; and Barbara Sylvester, Western Washington University.

Finally, I appreciate the excellent copyediting of the second edition by Patti Scott, the superb project management of Sheila Frank, and the professional, insightful guidance provided by a great developmental editor, Kate Scheinman.

01

Introduction

Technical writing is a fundamental skill for virtually anyone working in science and engineering—and that includes a broader range of people than just scientists and engineers. Most science and engineering activities must produce technical reports either on paper or in electronic form. Research, development, finance, manufacturing, and a host of technical commercial services rely on precisely written documents to communicate complex information to a wide range of audiences for many purposes. Technical writing is the means by which these documents are produced.

What Is Technical Writing?

To define what technical writing is, it might be useful to first clarify what it is not. Technical writing is not what one does out in the meadow under an elm tree; that is creative writing. Creative writing is, by and large, a pleasant activity. Technical writing, however, is tough, hard work. Technical writing is not what most people commonly do for fun or relaxation.

Imagine, if you will, a learned, artistic person sipping fine wine, eating imported cheese, and watching the sunset. This person sounds like a creative writer who, with paper and pen in hand, might reflect on the nature of time as the last rays of sunlight slowly disappear over the horizon. Our creative writer might even come up with a definition of *time* that goes something like this:

> *Time* is a river flowing from nowhere[1] through which everything and everyone move forward to meet their fate.

A creative, sensitive person sees an inspiring sunset, is moved to words, and writes the "river from nowhere" definition. Obviously, this approach is metaphorical. But what if someone inspired by that magnificent sunset were to write this definition of time?

Time is a convention of measurement based on the microwave spectral line emitted by cesium atoms with an atomic weight of 133 and an integral frequency of 9,192,631,770 hertz.

Perhaps not! What sort of individual would be emotionally moved by a sunset and then come up with microwave spectral lines and cesium 133? This would be a very strange person. The second definition of time is technical. It is designed to be objective, direct, and precise. Consequently, it lacks the emotional impact of the first definition because, as technical writing, it avoids the use of rich metaphors and figures of speech, substituting instead precise, empirical data.

The difference between the two definitions shows the fundamental distinction between technical writing and creative writing (and all the other kinds of writing that fall in between). Technical writing is precise, objective, direct, and clearly defined.

Abstraction   In linguistic terms, technical writing is writing that displays a relatively low level of abstraction. To clarify the concept of abstraction, consider the ladder of abstraction[2] in Figure 1.1. Here, various levels of abstraction are being used to refer to a 33,000-ohm, 1-watt carbon resistor. The lowest level of abstraction would be the resistor itself. However, we normally do not paste resistors into technical documents; so to be precise, we have to substitute something else, such as a photograph.

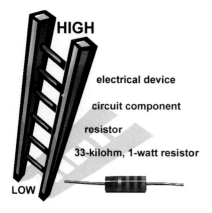

HIGH

electrical device

circuit component

resistor

33-kilohm, 1-watt resistor

LOW

Figure 1.1
Abstraction ladder.

That is why a photograph of the resistor has been placed at the bottom of the abstraction ladder— because that is the most concrete way available to refer to it in a document. This precision, of course, assumes that the audience knows what a resistor looks like. If not, the photograph can be pretty abstract, and for that matter, so can the actual resistor.

The next most concrete way of describing the resistor by using just words alone is to precisely label it—as we do when we give our newborn children distinctive names and IRS taxpayer identification numbers. In this case, the resistor is labeled a "33-kilohm, 1-watt carbon resistor."

As we move up the ladder of abstraction, the references become increasingly imprecise. At the next level, *resistor* could mean any circuit component that impedes the flow of current. At the subsequent level, *circuit component* could mean any electrical device in the circuit, such as a capacitor, inductor, diode, or, in this case, a resistor. Finally, at the top of the abstraction ladder, *electrical device* could be anything that is electrical, from a fluorescent light, to a computer keyboard, to a stereo system. Of course, *electrical device* also refers to that 33-kilohm, 1-watt carbon resistor.

The point is that as one moves up the ladder of abstraction, the symbols become less precise and increasingly vague. In effect, this increased abstraction gives the reader greater freedom to interpret meaning as he or she wants, and not necessarily as the writer intended. In creative writing, that is probably good; in technical writing, that is always bad. The goal of technical writing is to eliminate abstraction. Simply speaking, successful technical writing restricts the reader's freedom of interpretation so that only one meaning can be concluded—the meaning intended by the writer.

A reader might interpret that river flowing out of nowhere as any number of things based on his or her prior experience and emotional makeup. (For example, the author learned to water-ski on a river flowing out of nowhere—it is called the Potomac!) Unlike the river from nowhere, cesium 133 is cesium 133, and 9,192,631,770 hertz is 9,192,631,770 hertz, no matter what the reader's prior experience or emotional makeup. What makes the technical definition of *time* technical, then, is that it effectively restricts the reader's ability to abstract various meanings.

The potential for abstraction can never be totally eliminated, even by the best technical writing. The process of human perception presupposes some abstraction. Who knows? The reader might have flunked out of Chemistry 101 as an undergraduate 10 years ago because he could not find cesium in the periodic table of elements. Perhaps flunking chemistry caused him to lose his scholarship, which, in turn, led to his leaving the university. He was subsequently dumped by his girlfriend, the valedictorian of their high school and the one great love of his life. In despair, he became a bum. She, ironically, became a Nobel Laureate, probably for her work with cesium. So for this reader, the technical definition of *time* might have a connotative, emotionally charged meaning far apart from its precise, denotative function. It could happen; but in reality, it is certainly not likely.

What sets technical writing apart, then, is its precision. How it achieves this precision is, in fact, the art and craft of technical writing—an activity that involves definition and description; data and analysis; photographs, diagrams, and charts; and often specialized language. The goal of technical writing, then, is not to be creative or interesting; it is not to employ rich imagery or powerful metaphors. The goal of technical writing, first and foremost, is to communicate complex information clearly and precisely for the audience and the purpose at hand. Clarity and precision are the overriding goals for any technical writer, and understanding the audience and the purpose is the primary consideration for achieving those goals.

Audience and Purpose

The measure of how well a technical writer has written something comes down to two things: (1) how well the reader understands, precisely, the writer's intended meaning and (2) how well that understanding fulfills the intended purpose or the need at hand. Consequently, technical writing must be geared directly to its audience and purpose. Remember, there is always some specific requirement for technical writing: A scientist needs to write a proposal for a grant, a programmer needs to document a software package prior to distribution, or a lab chief needs to write a feasibility report for selecting new equipment.

Technical writing has to not only relate specifically to the purpose and situation at hand but also relate specifically to the reader, or audience, who will be using the document. In other words, the writer must consider the potential reader's knowledge, skill level, and specialization; and the writer has to fully respond to the needs of the reader in terms of the requirements of the situation.

To further clarify these concepts, let us briefly define a requirement and audience. Suppose that person watching the sunset who was writing about spectral line emissions and time has a problem with his social life. The problem is that he does not have one. By random chance, let us say he has finally met a young lady who has captured his heart. She, of course, scarcely knows he exists. He has decided to rectify the situation by writing her a note articulating his love for her. Coming up with the right words to win her heart is the requirement, and she is the audience.

The man has spent the day reading psychology journals on the concept of love and now believes he has the requisite knowledge to write about the subject. So that night, he pens a note to the woman of his dreams:

> Whenever I look into your eyes, I know that, from my perspective, I share with you a strong, interpersonal passion or enthusiasm statistically related at .05 or better to increased levels of self-disclosing behavior.

This writer may have the right idea, but he is doing the wrong kind of writing for his purpose and audience.

The same thing would be true of a social scientist preparing a scholarly paper on the occurrence patterns of erotic love among middle-aged women and men. Suppose this scientist writes the following:

> Love is the bond that holds humanity together and the rapier that rips it apart.

Certainly he or she could make this point, but given the purpose of the academic conference and the scholarly nature of the audience, this scientist probably would have to reword the passage to read something like this:

> In terms of interhuman social structures, love is an emotional quality that can exhibit adhesive or divisive properties or functions in response to outside stimuli.

This definition is technical writing; as such, it is better suited for a group of social scientists researching interpersonal communication patterns of subjects in experimental and control groups.

Report Writing and Audience

In technical writing, the audience and purpose are almost always well defined in advance—usually by your boss or teacher. The reason is that technical writing is basically report writing, meaning that it is writing commissioned by someone else for a specific purpose and audience. Normally, reports aim to share objective information with an interested, educated audience. Technical reports are simply reports on a technical subject that share this information in a precise way.

So, finally, to answer the question posed at the beginning of this chapter, technical writing can be described as follows:

- *Technical writing deals with technical information.* As the previous examples show, using technical writing for nontechnical purposes and situations is a bad idea. Technical writing is designed to deal with technical subjects.
- *Technical writing relies heavily on visuals.* As shown in the abstraction ladder of Figure 1.1, a photograph or diagram may be the least abstract, and therefore the most precise, way to communicate something to a technical audience. Visuals—

whether they are equations, photographs, tables, graphs, drawings, or charts—are powerful tools for providing a large amount of information effectively and efficiently. However, they almost always require interpretation or explanation in the text of the report, especially when the audience may not be familiar with them.

- *Technical writing uses numerical data to precisely describe quantity and direction.* In many cases, mathematical equations and values provide the real substance of a technical report.

- *Technical writing is accurate and well documented.* Generalized, unsupported assertions have no place in a technical paper. Conclusions, recommendations, and judgments are always based on clearly presented evidence or established expertise, and technical writing is always technically correct.

- *Technical writing is grammatically and stylistically correct.* This kind of correctness is far more than simply a matter of personal and corporate pride. Grammar and style often go to the heart of the author's credibility. In other words, if you write like an idiot, the reader may well perceive you as an idiot and your organization as a collection of idiots. Whether this perception is true may be irrelevant. Where technical writing is concerned, perception is often reality.

The following pages provide a straightforward, easy-to-follow tutorial on how to do technical writing. The author has tried not to take himself too seriously, because the goal is simply to get to the point and provide the necessary information in a way that is interesting, useful, and understandable. Here is a brief overview of the pages that follow.

- Chapter 2 focuses on ethics in technical writing and includes a general discussion of ethical considerations for technical writers, followed by specific discussions of plagiarism and image alteration.

- Chapters 3 through 5 explore the "nuts and bolts" of technical writing, including technical definition, mechanism description, and process description. The ability to define and describe is one of the primary skills needed for doing any kind of technical writing.

- Chapters 6 through 11 look at the most common types of technical writing documents: proposals, progress reports, feasibility and recommendation studies, laboratory and project reports, instructions and manuals, and research reports. Evolutionary descriptions and straightforward models are included for each type of document.

- Chapter 12 presents an easy-to-understand treatment of abstracts and executive summaries, which often accompany formal technical reports.

- Chapter 13 deals with grammar and style. This chapter does not provide the comprehensive treatment found in complete style manuals; rather, it focuses specifically on the most common grammar and style errors in technical reports today and shows how to fix these errors.

- Chapter 14 provides a no-nonsense guide to technical documentation, including the "why, when, and how" of documenting sources.

- Chapter 15 focuses on the design and use of visuals in technical documents and presentations. Topics include diagrams, charts, pictographs, schematics, and photographs. A discussion of photograph alteration and image compositing is also included here (the ethical dimensions of this topic are treated in Chapter 2).

- Chapter 16 explores some of the major differences between traditional and electronic publishing and reviews the major file formats used in electronic publishing today.

- Chapter 17 shows how to put together technical presentations and presents a simple model for organizing technical briefings.
- Chapter 18 deals with resumes, cover letters, and job interviews for engineers and scientists; and it provides an effective approach for seeking technical positions in the real world.
- Chapter 19 looks at team writing and describes how two or more people can work together to produce technical documents. The chapter also highlights the differences between team writing in student and professional environments.

Notes

1. The idea of a "river from nowhere" came from the Steve Winwood song "The Finer Things," which includes a reference to time as a "river rolling into nowhere."

2. The idea of an abstraction ladder was borrowed from S. I. Hayakawa, *Language in Thought and Action* (New York: Harcourt, Brace and Company, 1949), p. 160.

Ethical Considerations

Why is ethics the focus of a lead chapter in this book? It is here because of increasing emphasis being placed on ethics in technical writing, as well as in engineering and the sciences as a whole. Technology and the communication of technical ideas represent powerful tools for promoting good or evil in society—and ethics is an important backdrop for all the material provided in the chapters that follow.

Humanity has been struggling for a long time over exactly what constitutes good and evil, and ethical behavior. It is unlikely that this chapter in a technical writing book will provide the definitive answer that will end this timeless struggle. The hope, however, is that this chapter will provide a minimal awareness of the ethical dimensions of technical writing and some sensitivity to the real issues that exist within these dimensions. Additionally, this chapter will discuss two key areas of ethics in technical writing—plagiarism and image alteration—where the power of modern digital technology has heightened ethical concern.

For a more extensive treatment of ethics, not just in technical writing but also in the broader context of engineering and science, many excellent books exist, including Martin and Schinzinger's *Ethics in Engineering*.[1]

What Are Ethics in Technical Writing?

The first thing we need to develop is a working definition of ethics in technical writing. What exactly are ethics for a technical writer? For the purposes of this book, the answer is relatively simple. Collectively, *ethics* is a set of rules and standards for using communication skills and resources with the intention of doing good. Ethical behavior for technical writers, then, involves their moral duty and obligation to apply the power of technical communications with the purpose of doing worthy things. But what do *good* and *worthy* mean?

You probably already know what you think they mean, but not everyone would agree with you. As will be discussed in Chapter 16, the global technical writing community, especially with the advent of electronic publishing and the World Wide Web, is composed of many disparate cultures. To some extent, concepts of good and bad, and ethical and unethical, are culturally relative. That does not mean we can just ignore the issue. But we need to approach the matter realistically in the context of the pluralistic environment in which we function. That is what this chapter attempts to do.

Begin by considering the following situation:

A middle-aged technical writer writes an irreverent, humorous technical writing book using, as its examples, reports on fictitious technologies based on real theory. His laboratory report chapter deals with a fictional, high-power transmitting tube designated the *16XL1000000,* his research report chapter describes a quantum computing processor known as the *QuantumCPU,* and his proposal and progress report chapters are built around scientific analysis of a figure skating jump called the *QuadFINKEL.* For the sake of argument, assume that the book is a big hit, the royalties pour in, and the author deposits the checks and goes out to live the good life.

Now consider this situation:

> A middle-aged technical writer writes a professional proposal from a fake corporation he names after himself. He invites individual investment of funds in developing and marketing the fake company's fictitious product line of high-technology devices. He cites invented test data and technical research information demonstrating the efficacy of these products. In fact, he develops a professional, full-color brochure complete with charts and diagrams—and promises huge returns on investment. He mails this brochure to thousands of nursing home residents across the country. When many of these people send him their life savings, he deposits the checks and goes out to live the good life.

In both situations a technical writer has used his knowledge and resources to develop published materials around fictitious technologies. In both situations he has embellished the capabilities of these technologies with professionally developed diagrams, charts, and data. In both situations he makes money from the published product.

In ethical terms, how do these situations differ? That is an important question; the difference gets at what makes the behavior of technical writers ethical or unethical. Before we get to this difference, however, we need to look briefly at the kinds of ethical constructs that are traditionally used in technical writing and the implications of these constructs for technical writers:

- Technical writers must be accurate in their work. Either technical writers must be precisely correct at all times, or they are unethical.
- Technical writers must be honest in their work. Technical writers who write untruths are unethical.

- Technical writers must always honor their obligations. Technical writers who do not produce the documents and other materials they are responsible for producing within the agreed-upon time frame are unethical.
- Technical writers must not substitute speculation for fact. Technical writers who do not clearly separate opinion from accepted truth are unethical.
- Technical writers must not hide truth with ambiguity. Technical writers who play down facts that would be contrary to the theses of their reports are unethical.
- Technical writers must not use the ideas of others without giving proper credit. Technical writers who fail to document the sources of all nonoriginal ideas, except for common knowledge, are unethical.
- Technical writers must not violate copyright laws. Technical writers who fail to document the use of copyrighted materials when used with permission or under "fair use" are unethical. Additionally, technical writers who use any copyrighted materials without permission when these uses are not covered by fair use are unethical whether the materials are documented or not.
- Technical writers must not lie with statistics. Technical writers who manipulate data or graphical representations of data, use inappropriate or improper statistical tests, or employ loaded statistical samples are unethical.
- Technical writers must not inject personal bias into their reports. Technical writers who are less than objective in everything they write are unethical.

What a list! These rules are clearly well intentioned, and they provide generally useful guidelines for writers. But there is a real problem with rules like these: They miss the mark where ethics are concerned. For example, being accurate and precise in technical writing is not the real ethical issue here. Intending to be good and to do good is the real issue. Of course, it is true that in technical writing "being good and doing good" usually mean being accurate and precise, but not always.

In the first situation, as the author in question, I obviously do not believe that being accurate and precise is necessary to achieving my pedagogical goals for this book. In this case, I believe that just the opposite is true. The freedom to invent obviously fictitious technical writing examples supports the book's educational objectives. I am being inaccurate, but I am not being deceptive. For example, in Chapter 5, the exercise deals with the QuadFINKEL figure skating jump, an athletic achievement so technically demanding that anyone landing it always receives the gold medal. Do you suppose anyone really believes that a QuadFINKEL exists, or that the author can even stand up on a pair of figure skates? Of course not.

I would be deceptive, and therefore unethical, only if I asked that you, the reader, truly believe these examples. My clear intent is not to mislead; rather, it is to use the fictitious and humorous examples to help students learn—and I believe learning is still considered a "good" thing to do. However, I wouldn't do well with that list of rules, would I? These rules do not get at ethics; rather, they get at behavior that is often correlated with ethics.

In the second example, the guy trying to rip off nursing home residents is obviously an unethical toad (he is so bad, in fact, that this statement disparages toads!). He clearly intends to use his technical communications skills and resources to deceive. To make matters worse, he is also targeting a vulnerable group of people for this scam—the elderly, who are least able to afford this kind of attack. So unlike in the first situation, this guy is falsifying information not to do good, but to do bad. That is why the technical writer in the second example is absolutely unethical (not to mention being guilty of violating numerous local, state, and federal laws).

Figure 2.1 presents a model for ethics in technical writing. Clearly, it does not represent the big breakthrough in the philosophy of ethics for which humanity has been waiting so long. But it does help make the point and clarify the issue. The rules may contribute to ethics and our understanding of what is involved in ethics, but they are not what constitute ethics!

Finally, ethics in technical writing, to some extent, must be relative to societal values. If, on one hand, you use technical communications to deceive your reader with the goal of doing bad things (defining *bad* by the standards of your society and perhaps by the standards of civilization as a whole), you are unethical. If, on the other hand,

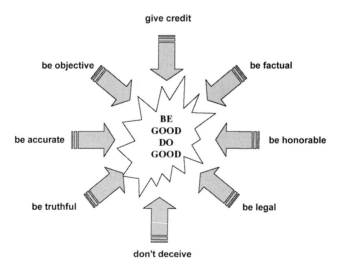

Figure 2.1
Ethics model for technical writing.

you use technical communications with the intention of doing good things (defining *good* in the same manner), you are ethical.

Plagiarism

Plagiarism is basically an act of theft in which you steal another person's idea, or his or her expression of an idea, and then represent it as your own. Using someone's ideas without acknowledging the source constitutes academic misconduct (unless the ideas are *common knowledge*— see Chapter 14). You are also guilty of plagiarism if you steal someone's *expression* of an idea—that is, the words, images, sounds, photographs, etc., produced by another.

Plagiarism is a growing problem primarily as a result of the Internet. Not that many years ago, plagiarism was controlled somewhat by the difficulty of plagiarizing and the risk of getting caught. If students wanted to steal material that they could represent as theirs to fulfill a class assignment, they usually had to go to the library and look through a limited collection of materials probably known to their teachers. Students might even have to pay to use a copy machine or endure the drudgery of transcribing the material manually. When someone plagiarized another's work, the act itself involved time and effort and included the ever-present risk of detection. And in some cases, plagiarism actually cost more in time and money than it was worth.

Those days are over! Now the Internet has extraordinary search capabilities that provide a plagiarist with quick access to literally thousands of sources and documents. With the press of a key, any number of files can be transferred to the plagiarist's computer in absolute privacy and then easily converted to a word processing file. This file can be searched and modified to remove any reference to the original author. Then the document can be adapted to course requirements and submitted for credit.

Not only is the risk of detection virtually non-existent during the act of piracy, but also the risk of detection after the act, while not insignificant, is relatively low. Most online documents are otherwise unpublished—many are written by unknown authors—and many of these documents are no more than recycled student papers. Additionally, commercial "paper mill" websites exist that provide papers to students on a wide range of topics for a fee. As one site advertises, if students are willing to spend a few bucks, they can "download their workload."[2]

Another factor at work involves compromises in the ethical imperative not to cheat. Clearly, students of pre-Internet days, who would not have plagiarized solely because they feared detection or did not want to go to the trouble, would readily cheat today, using current Internet resources. This results in larger numbers of cheaters today. For some struggling with the ethics of plagiarism, the fact that "everyone" seems to be doing it somehow makes doing it more acceptable. It is the same phenomenon we see with otherwise honest, decent people readily exchanging music files or software in digital format over peer-to-peer Internet connections.

Additionally, for some, stealing files from the Internet may seem less wrong than, say, copying pages from a book. Many students see information available on the Internet as public property that is theirs for the taking—not unlike old furniture a neighbor has put out at the curb on trash collection day. The ephemeral nature of files on the Internet, the lack of material tangibility, the speed and impunity with which this material can be located and reproduced, and the ease with which it can be used in private—all contribute to the perception that somehow stealing a file from the Internet is less wrong than copying from a book or journal in the library.

Fortunately, most students are not ethically challenged—and for those who are, the same technology that gives them the means to cheat also provides teachers with the power to detect. The same search engines that help plagiarists find materials to steal can also help teachers find the same materials when plagiarism is suspected. Counterplagiarism websites such as www.turnitin.com and www.plagiarism.com also provide effective resources to help teachers identify plagiarism.

The bottom line is this: In technical writing, as in all intellectual pursuits, stealing ideas or the expression of ideas from others is not ethical and is absolutely unacceptable. It does not matter that "everyone" seems to be doing it, or that online resources make locating and using such material a trivial task. If you use the ideas or the expression of ideas (words, images, sounds, photographs, etc.) of others without acknowledging the source, you are stealing from the author, you are stealing from those students who are doing their own work, and you are stealing from yourself by turning what could be a valuable educational experience into an exercise in thievery. Also, if you commit plagiarism, especially in the professional world, you might find yourself on the wrong side of copyright, patent, trade secret, or trademark protections. That could subject you and your employer to severe criminal and civil sanctions, which, at the very least, would not be a career enhancement.

Learn to do it right the first time while in school. Document all sources with citations or notes at the point in the paper where the materials are used, and include a complete list of sources, usually at the end of the paper. See Chapter 14 for a more complete discussion of documentation and documentation requirements.

Image Alteration and Ethics	Without a doubt, digital image processing lends itself to abuse and misuse. The relative ease, availability, and effectiveness of hardware and software, combined with the inherent credibility of photographic images, provide a powerful tool for misrepresenting reality. As previously discussed, modern technology provides almost anyone with the power to steal or plagiarize intellectual property in digital form. Additionally, this same technology provides technical writers with the means to modify material with the intent of misleading the reader.

This is especially true in the case of "doctored" photographs. Technical writing always must be accurate and correct, and serious ethical concerns exist regarding accuracy and correctness when images are edited. Chapter 15 provides a discussion of image alteration and makes the point that modifying images is ethical or even desirable in technical writing when these modifications serve to *clarify* or *complement*—but not when they are used to *misrepresent* or *deceive*. If the image alteration has fundamentally changed the information it represents for the purpose at hand, that fact must be acknowledged. To avoid any ethical questions when you are using such altered images, always indicate to your reader that the images are altered. The easiest way is to acknowledge in the image's title and caption that it is, in fact, an edited image and then to briefly describe the nature and purpose of the alteration.

Plagiarism	**Exercises**

Situation 1

You are enrolled in a technical writing course and have been assigned the task of writing a process description on the operation of a four-stroke, internal combustion engine. A close friend wrote a similar description for another technical writing course at a different university last year and made an "A." She offers you her paper as a gift, along with the original word processing file. She tells you to change the name, print it again, and turn it in. Since the original author has given you the paper and you now own it, is it plagiarism for you to do as she asks? After all, she argues, you are not stealing anything—the paper is now yours!

Situation 2

What if you actually wrote the paper that your close friend turned in under her name a year ago and for which she received an A. Now you are in a technical writing class at another university and have been given the same assignment. You still have the word processing file for her paper on your hard drive. Since you were the original author, is it plagiarism if you change her name to yours and turn in the paper for credit? If so, do you even need her permission?

Image Compositing

Photo 2.1*a–d* shows three separate source images taken at the Kilauea volcano, Hawaii, and a composite image developed from them. Photo 2.1*a* shows the author's wife, Phyllis, standing on the lava flow with many people in the distance behind her. Photo 2.1*b* shows the author standing on another part of the lava flow with a young lady visible behind him (no, he did not know the young

Photo 2.1 (a)   Photo 2.1 (b)

Photo 2.1 (c)

lady). Photo 2.1*c* shows a sign warning of the danger of lava bench collapse on the coast, a location some distance from where the first two images were taken.

Photo 2.1*d* combines the three source images in a way that places the author and his wife together, removes all the people in the background including the young lady behind the author, and inserts the warning sign in the foreground. The author and his wife were never standing together as shown; they were never far from other people; and they were a safe distance from the lava bench that could collapse at any time.

For the following three hypothetical situations, is the composite image ethical? For each situation, if you believe the image is not ethical, what can you do to make it ethical?

- *Situation 1:* The author and his wife wanted to be pictured together on the volcano so that they could complete their scrapbook of their trip to Kilauea, but the picture of them together did not turn out.
- *Situation 2:* The author and his wife have been obsessed with impressing their friends and believe the composite photograph might achieve that purpose.

Photo 2.1 (d)

Photo 2.1
Kilauea lava flow, composite image.

- *Situation 3:* The author and his wife have applied for elite status in the Intrepid Explorer's Society and need the composite photograph as evidence that they had gone where no one else would go.

Notes

1. Mike W. Martin and Roland Schinzinger, *Ethics in Engineering,* 3rd ed. New York: The McGraw-Hill Companies, 1996.
2. The idea of "download their workload" came from the slogan "download your workload," Internet: www.schoolsucks.com.

03
00011

Technical Definition

Virtually any kind of technical writing includes one or more technical definitions. Consequently, a technical writer must be able to define terms, irrespective of whether these terms refer to mechanisms or processes.

What Is a Technical Definition?

In technical writing, *definition* is the process by which one assigns a precise meaning to a term. To define a term, it must be placed into a classification and then differentiated from other terms in that same classification. Technical definitions are relatively easy to write, except for some pitfalls that will be addressed later. The format for a technical definition is straightforward and works like this:

Term = **Classification** + Differentiation

For example, if a writer were to define the stall condition that an airplane experiences when it loses lift, he or she could start with the term *stall,* add a classification—flight condition—and then differentiate it from all other flight conditions, in this case, by a stall's unique characteristics. The definition might read something like this:

A *stall* is a **flight condition** in which the lift produced becomes less than the weight of the airplane, and the airplane stops flying.

That seems simple enough; but what happens when a term, such as *stall,* has multiple definitions

25

in many contexts? In such cases, it may be necessary to add a qualifier in front of the definition statement to supply the necessary context. The qualifier is important when the general context for a definition needs to be established up front. If the context is known or is obvious, a qualifier is unnecessary. For example, in an aeronautics study on aircraft wing design, the context of *stall* is obvious. It is clear that *stall* in this case has to do more with the loss of lift than with, say, a single compartment for an animal in a barn.

When a context is needed, the format for the definition is

$$\text{(Qualifier +)}\ \textit{Term}\ =\ \textbf{Classification}\ +\ \text{Differentiation}$$

Note: The parentheses, italic, and boldface have been added here and earlier for clarity.

Look at Figure 3.1, which provides three definitions of the same term in different contexts. Notice how the term *stall* has three totally different meanings depending on the context, and how each definition begins with a qualifier that makes the context clear from the start.

In the first example, *stall* refers to what happens when an airplane does not go fast enough to stay in the air. Pilots routinely stall their airplanes right above the runway when landing them, in which case stalling is good. Sometimes, however, they stall them inadvertently in situations when they really want the planes to keep flying, in which case stalling is bad.

In the second example, *stall* refers to a car that has suddenly stopped running. Normally, this condition happens only in the middle of heavy traffic, in bad weather, or when you are desperately trying to make a flight for which you have unrefundable tickets.

The third example of stall relates to social dating behavior. The author had a great deal of

Qualifier   Term   Class

 (**In aeronautics**), a *stall* is a **flight condition** in which the lift produced is less than the weight of the airplane.

 (**In driving**), a *stall* is an **operating condition** in which a sudden and unexpected loss of power occurs.

(**In dating behavior**), a *stall* is an **interpersonal maneuver** used by one party to discourage the unwanted advances of another.

Figure 3.1
Multiple contexts and qualifiers.

experience with this one during his undergraduate years. The typical "goodnight kiss" stall was for his date to fill her mouth with bubble gum and start chewing.

Often, the most difficult part of writing a technical definition lies in determining the proper classification for the term. The class should be a general category in which the term fits, but it cannot be too general. For example, consider that 33-kilohm, 1-watt carbon resistor from the abstraction ladder in Chapter 1. In the following sentence, the term is defined by using the very generic classification of *device*.

Classifications and Classes

> The *33-kilohm, 1-watt carbon resistor* is a device that impedes the flow of electric current.

The problem with this classification is that *device* could mean all kinds of different things, most of which have nothing to do with circuit components; consequently, its inclusion does not really help specify the meaning of the term. By changing *device* to *circuit component,* however, the meaning can be narrowed considerably for the reader, even before it is differentiated.

One trick for classifying a term is to build an abstraction ladder for the term and then move up one or two "rungs" above the term. In the original abstraction ladder, the movement was as shown in Figure 3.2. In this case, moving up one rung—from the term *33-kilohm, 1-watt carbon resistor* to the term *resistor*—is not an option because the classification would be derived from the original term. That would yield the following circular definition:

> The *33-kilohm, 1-watt carbon resistor* is a <u>resistor</u> that impedes the flow of electric current.

Because the 33-kilohm, 1-watt resistor is obviously a resistor, this classification does not help define the term. In fact, it contributes nothing other than useless circularity. If possible, try to find a classification that is not derived from the term. In this case, the easiest solution is to just move up the abstraction ladder one more rung. *Circuit component* helps to define the term and has no circularity with the term.

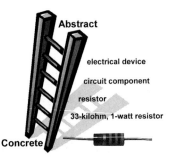

Figure 3.2
Abstraction ladder.

The next step in defining the term is to differentiate it from all the other members of the class. Differentiation involves narrowing the meaning of the term to just one possibility within the class. Clearly, it would be easier to narrow the class of circuit components to a particular resistor than it would be to narrow the class of device to a particular resistor. There are all kinds of devices in the world and relatively few circuit components.

Differentiation

The class "circuit components," however, still contains many possibilities: capacitors, diodes, switches, potentiometers, inductors, transistors, and IC chips, to name a few. In this case, a good approach is to focus on the function of a resistor, which is to impede the flow of electric current, and to use that function to differentiate the class. Doing so yields the following definition:

> The *33-kilohm, 1-watt resistor* is a circuit component that impedes the flow of electric current.

In writing technical definitions, it is easy to do something that is "not good." For example, when we are defining the common computer term *hard drive,* it is probably best to add a qualifier unless the context of computing is obvious. (There are other kinds of hard drives, such as one through the Mojave Desert without air conditioning in July.) So how about this definition?

Avoiding Mistakes

> In computing, a *hard drive* is an input/output device for the nonvolatile storage and retrieval of data.

Think about this one for a minute. If the reader does not know what a hard drive is, what is the likelihood that the reader will know what an input/output device is, much less the meaning of *nonvolatile storage and retrieval?*

How about this definition of a high-power transmitting tube?

In broadcast engineering, the *16XL1000000* is a high-power tetrode capable of producing more than a megawatt of RF.

This definition is fine as long as you are sure your audience will know what *tetrode* or *megawatt* or even *RF* refers to? An expert technical audience certainly would, but what about a less technical audience—perhaps a nontechnical manager who controls your funding? For a less technical audience, we might rewrite this definition as follows:

In broadcast engineering, the *16XL1000000* is a high-power transmitting tube capable of producing a radio or television signal with more than a million watts of power.

Notice that there is no mention of *tetrode* or even that the tube has four elements. A nontechnical audience would not know (or probably even care) about such things as cathodes, grids, and plates. Also, *RF* has been changed to *radio or television signal* and *megawatt* has been changed to *million watts of power.*

Extensions
You must always consider the reader's knowledge and skill level when defining terms and concepts. However, at times you may have no choice but to violate the principle to achieve your goal. Sometimes a simple term or concept that the audience will understand is just not available. In fact, sometimes the only thing to do is to define the term as best you can, then add extensions to your definition to clarify your meaning.

Extensions can take many forms. Here are a few of the most common types of extensions, along with examples of how they might relate to the expert-level definition of the 16XL1000000 discussed earlier. In the following examples, the original definition is included in brackets.

Further Definition

Use *further definition* when you need to define terms you used in your original definition.

[In broadcast engineering, the *16XL1000000* is a high-power tetrode capable of producing more than a megawatt of RF power.] *Tetrodes* are tubes with four elements, including a cathode, control grid, screen grid, and plate.

Comparison and Contrast

Use *comparison and contrast* when you need to show differences or similarities.

[In broadcast engineering, the *16XL1000000* is a high-power tetrode capable of producing more than a megawatt of RF power.] This fictitious tube is similar in design and function to the real-life, but lower-powered, Eimac 4CM400,000A, which also uses a ceramic-metal structure and water-vapor-cooled, thoriated-tungsten mesh filament.

Classification

Use *classification* when you need to organize information into categories.

[In broadcast engineering, the *16XL1000000* is a high-power tetrode capable of producing more than a megawatt of RF power.] The tube is classified as a metal-ceramic, coaxial, power-grid transmitting tube based on its material construction and the operational profile of its elements.

Cause and Effect

Use *cause and effect* when you need to demonstrate why something happens or when you need to trace results.

[In broadcast engineering, the *16XL1000000* is a high-power tetrode capable of producing more than a megawatt of RF power.] The tube functions by emitting electrons from a filament cathode through a grid structure to a positively charged plate. A control grid regulates the flow of electrons much as a valve does, while a screen grid provides necessary isolation between input and output elements.

Process

Use *process* when you need to list the steps of a procedure.

[In broadcast engineering, the *16XL1000000* is a high-power tetrode capable of producing more than a megawatt of RF power.] Preventing self-oscillation in the tube requires the application of negative feedback through the process of neutralization. The output signal is sampled from the plate circuit, coupled through a neutralizing capacitor, and then applied out of phase through a center-tapped input transformer.

Exemplification

Use *exemplification* when you need to give real or analogous examples.

[In broadcast engineering, the *16XL1000000* is a high-power tetrode capable of producing more than a megawatt of RF power.] The tube is commonly used as a class C power amplifier in military or broadcast service (shortwave, FM, or VHF TV) between 3.9 megahertz and 150 megahertz. It is also used in class AB_1 audio service as a high-level amplitude modulator.

Etymology

Use *etymology* to show the linguistic genesis of the term.

[In broadcast engineering, the *16XL1000000* is a high-power tetrode capable of producing more than a megawatt of RF power.] The term *16XL1000000* comes from the tube's operational parameters, where 16 refers to the filament voltage, XL refers to its designation as a high-power transmitting tube, and 1000000 indicates its rated level of RF power.

In some situations you may need to trade off **Required** desired precision in your definition to achieve the **Imprecision** required level of communication. At times it is foolish to attempt to achieve expert-level precision with an uninformed audience.

Consider the following two definitions of a black hole—the astrophysical phenomenon that is supposed to exist somewhere in space.

In astrophysics, a *black hole* is a set of events from which it was not possible to escape to a large distance. A black hole gets its name from its boundary, called an *event horizon,* which is formed by the paths in space-time of rays of light that just fail to get away, hovering instead forever on the edge and, consequently, moving on paths parallel to or away from one another.

In astrophysics, a *black hole* is a collapsed neutron star whose gravity is so great that even light cannot escape. Although fusion reactions within this collapsed star still may emit brilliant rays of light, when it is viewed from the outside, the black hole appears to be a totally dark void in space.

The first definition functions at the expert level. For theoretical physicists, it provides a precise and accurate description of a black hole and thus is appropriate for their needs. But for the less informed, reading it represents a mind-twisting experience that, in many cases, can leave readers more confused about the term than they were before they read it.

The second definition functions at the level of the average reader. It is not nearly as precise or accurate as the first definition, but it communicates the basic gist of what constitutes a black hole in space.

The problem is that to be precisely correct and absolutely accurate, you would have to function at a level where you cannot effectively communicate with the average reader. If you were writing for the average reader, you would have to make a decision here: Either be absolutely correct and communicate less effectively, or be less than absolutely correct and communicate more effectively. Since your goal is effective communication, your decision is obvious.

A Word about Defining Specifications and Standards

Defining specifications and standards is a specialized activity. Such documents can take many different forms, depending on what areas of engineering and science are involved and whether the documents are designed to meet commercial, industrial, or government standards. Writing these kinds of definitions is far more complex than simply defining terms.

Specification documents precisely state particulars, including requirements, designs, implementations, and testing. In engineering and science, specifications normally involve goods and services being developed under some type of contractual obligation. The specification precisely defines the quality of work and performance standards required by the contract. In addition, as will be shown in Chapter 4, specifications can describe mechanisms in terms of their physical and functional attributes.

Specifications are also what technical standards are made of. *Standards* are accepted or established methods, measures, or designs for accomplishing specific tasks. Standards exist for everything from data transfer protocols and cable connectors to air conditioning coolants and drinking water.[1] Specifications used in standards are detailed and exacting. The following excerpt from the IEEE 1394 Open Host Controller Interface Specification is a good example:

> [IEEE] 1394 requires certain 1394 bus management resource registers be accessible only via "quadlet read"

and "quadlet lock" (compare-and-swap_transactions), otherwise ack_type_error shall be sent. Those special bus management resource registers are implemented internal to the 1394 Open Host Controller to allow atomic compare-and-swap access from either the host system or from the 1394 bus.[2]

If the specification is required under U.S. government contract, it must contain certain information about the goods and services involved and the various standards that apply. Government specifications often require the following:

- Precise definitions and descriptions of the scope of the project.
- Any documentation the contractor must furnish, along with the formats for those documents.
- Specific performance characteristics of any required product, along with necessary testing, including procedures and equipment, to verify that the goods or services meet the specified requirement.
- Exact descriptions of the deliverables of the contract, including all goods and services, and the dates and times by which these products will be provided.
- Contractor notes, records, and other research and production materials.

Writing specifications is a demanding task normally accomplished by experienced engineers, scientists, and project managers with a solid knowledge of all applicable standards. Standards often are developed by committees that are composed of legal, managerial, and technical experts.

- Have I fully analyzed the purpose of my report, and do I understand the skill and knowledge level of the audience? **Definition Checklist**

- Have I defined the term by first classifying it in a way that adds precision and understanding for my audience and that serves my purpose?
- Have I differentiated this classification to distinguish this term from other members of its class?
- Have I determined whether the context is clear and, if not, whether it is critical to the definition? If the context is unclear and critical, have I used a qualifier before the term?
- Have I avoided defining a term with the same term?
- Have I avoided using terms that themselves need to be defined? If not, have I explained these terms?
- Have I chosen extensions to my definitions that are appropriate for my audience and purpose?
- Have I compromised my fundamental purpose (communicating with the reader) by including inappropriate or irrelevant information or precision?

Exercise Read each of the following definitions and try to determine the context, audience level, accuracy, and purpose for which they were written.

- A *resistor* is a small electronic part that reduces the amount of electricity flowing through a circuit.
- A *resistor* is a circuit component that converts electrical energy to thermal energy and, in the process, determines the current produced by a given difference of potential.
- *Resonance* is a systemic condition in which small amplitudes of a periodic agent produce large amplitudes of oscillation or vibration.
- *Resonance* is a natural means of amplification that makes a musician's horn sound louder.
- *Ionization* is the electrostatic process by which a neutral atom or molecule loses or gains electrons, thereby acquiring a net charge.

- *Ionization* is the phenomenon that creates lightning in thunderstorms.
- *Ergonomics* is the field of study by which we make machines easier to use.
- *Ergonomics* is the systematic consideration of physical, psychological, and social characteristics of human beings in the design of tools and equipment, the workplace, and the job itself.

Notes

1. For an updated, comprehensive listing of standards, see "CFS Standards Document Library on the World Wide Web." Internet: http://www-library.itsi.disa.mil/by_org.html, March 23, 1999. Some of the more commonly used standards on this website include those by the American National Standards Institute (ANSI) at http://www-library.itsi.disa.mil/org/ansi_std.html; Department of Defense Standards (DOD-STD) at http://www-library.itsi.disa.mil/org/dod_std.html; Institute of Electrical and Electronic Engineers (IEEE) at http://www-library.itsi.disa.mil/org/ieee_std.html; International Organization for Standardization (ISO) at http://www-library.itsi.disa.mil/org/iso_std.html; Military Standard (MIL-STD) below 2045 at http://www-library.itsi.disa.mil/org/mil_stdb.html; Military Standard (MIL-STD) 2045 and up at http://www-library.itsi.disa.mil/org/mil_std.html; and Telecommunications Industry/Electronic Industries Association (TIA/EIA) at http://www-library.itsi.disa.mil/org/tia_eia.html.

2. Apple Computer, Inc., Compaq Computer Corporation, Intel Corporation, Microsoft Corporation, National Semiconductor Corporation, Sun Microsystems, Inc., and Texas Instruments, Inc., "1394 Open Host Controller Interface Specification," Release 1.00, p. 38. (file=ohcir100.pdf). Internet: http://inanna.ecs.soton.ac.uk/~swh/mLAN/ohcir100.pdf.

CHAPTER FOUR

Description of a Mechanism

Technology involves physical devices called *mechanisms*. Being able to describe mechanisms precisely and accurately, and at a level and in a way that the reader needs and can understand, is perhaps the most essential component skill of writing technical reports. This skill is particularly important for those producing documents involving specifications and instructions.

What Is a Mechanism Description?

Mechanism descriptions are precise portrayals of material devices with two or more parts that function together to do something. Mechanisms can range in complexity from circuit components and mechanical fasteners, to supercomputers and space shuttles, to electric train sets a parent has to put together on Christmas morning.

In Outline 4.1, notice that the primary focus in writing a mechanism description is on the physical characteristics or attributes of a device and its parts. These documents are built around precise descriptions of size, shape, color, finish, texture, and material, and they can be written for a wide range of audience skills. Such descriptions also normally include figures, diagrams, or photographs that help to clarify the word description.

A more specialized mechanism description, the *functional mechanism description,* describes a mechanism with key physical attributes related to operating parameters. This type of description, which often includes specific operational standards, can be used to provide specifications for the

39

mechanism. Technical specifications normally are written for an expert audience who actually needs to design, maintain, use, or evaluate the mechanism.

Example of a Outline 4.1 provides a model for a general mech-
Mechanism anism description. To show how it works, we will
Description use it to describe a relatively simple mechanism: the 33-kilohm, 1-watt carbon resistor discussed

---

Outline 4.1 Description of a Mechanism

Introduction
- Define the mechanism with a technical definition (see Chapter 3), and add extensions to discuss any theory or principles necessary for the reader to understand what you are saying. Always make sure you add only what the reader needs for the purpose at hand. If the reader does not need any theory or operating principles, do not provide any.
- Describe the mechanism's overall function or purpose.
- Describe the mechanism's overall appearance in terms of its shape, color, material, finish, texture, and size.
- List the mechanism's parts in the order in which you plan to describe them.

Discussion
- Part 1
  - Define the first part with a technical definition, adding extensions as needed to deal with theory or operating principles.
  - Describe the part's overall function or purpose.
  - Describe the part's shape, color, material, finish, texture, and size (as well as any other physical attributes appropriate for the mechanism and its function), using precise measures and descriptors. Also, be sure to use figures, diagrams, and photographs as necessary.
  - Transition from this part to the next part.
- Parts 2–n
  - For each remaining part, repeat the pattern of defining, describing, and transitioning established for Part 1.

Conclusion
- Briefly summarize the mechanism's function and relist the parts described.
- Give a sense of finality to the paper.

in Chapters 1 and 3. The resistor is a relatively simple mechanism; most mechanisms are much more complex. Assume, for the purpose of illustration, that this description is for the average technical reader who requires only a general description of the resistor.

Introduction

Following the outline, first introduce the mechanism with a technical definition and extensions that describe its overall function and purpose:

> The 33-kilohm, 1-watt carbon resistor is a circuit component that impedes the flow of electric current.

Next, use extensions to discuss any theory or operating principles necessary for the reader to understand the description. Be careful here; think about what you are doing, for whom you are doing it, and why you are doing it. Do not lose sight of the reader's knowledge and skill level or forget the purpose of the report. The goal is not to show how smart you are, but rather to communicate the information. So, when you are discussing theory in a mechanism description, a good rule is to do what is necessary, but only what is necessary.

For example, consider this theoretical discussion:

> The resistor impedes the movement of free electrons, thereby generating a thermal response depending on temperature, cross section, and length of the resistive element. The resulting resistance is measured in ohms; a *resistor* is defined as having 1 ohm of resistance when an applied electromotive force of 1 volt causes a current of 1 ampere to flow. This resistor has 33 kilohms, meaning that when 1 volt is applied, a current of .00003 ampere would flow. In addition, the square of the current flowing in amperes, times the resistance in ohms, determines the power dissipated in watts. This particular resistor can safely and continuously dissipate 1 watt of electrical energy as heat.

This discussion might be fine for readers with more specific technical purposes, but not for the average person who needs only a general description. *Remember,* the goal here is to write a general, informative mechanism description. The purpose is not to provide a description of how a resistor works or to show how to measure its effect in a circuit. In addition, if this theoretical discussion were to be used for the intended audience, several additional concepts (such as current, voltage, and power) would have to be defined and discussed.

Given the reader's skill level and purpose, the following two sentences might provide a simplified theoretical discussion that is more appropriate:

> The resistor impedes the flow of current by converting a portion of the electrical energy flowing through it to thermal energy, or heat. This particular resistor can safely convert 1 watt of electrical energy into heat.

Notice that this description does not get into Ohm's law. (It also does not explain that if I placed 1,000,000 volts across this resistor, it would draw 30.3 amperes of current, resulting in the dissipation of more than 30 million watts of power, thereby generating an inferno that would destroy my house and probably the entire neighborhood.) This information is not accurate (the resistor would vaporize instantly); and if it were, it is not relevant to describing the mechanism. It might be relevant to describing the operation of the mechanism (or, if accurate, in answering the civil suits brought by the neighbors), but that is not the purpose here.

The next step in the introduction is to describe, in general terms, the mechanism's overall appearance—its shape, color, material, finish, texture, and size. In this case, one could describe the device as follows:

The 33-kilohm, 1-watt carbon resistor looks like a small cylinder with wire leads extending from each end. The cylinder's surface is composed of smooth, brown plastic with a shiny finish. Four equally spaced color bands (three orange, one gold) circumscribe the cylinder, starting at one end.

Finally, to complete the introduction, list the mechanism's parts in the order in which they will be described. This listing organizes the remainder of the mechanism description. Consequently, the decision regarding the order in which to list the mechanism's parts is not trivial. It effectively determines the structure for the rest of the mechanism description.

There are two ways to order the parts: spatially and functionally. Using a spatial organization, you can move from left to right, or top to bottom, or inside out, or outside in. Using a functional approach, you can order the parts in terms of how the parts function with one another. Functionally, for example, you could start with the leads, which connect the circuit to the carbon element, which is protected by the casing, and around which the color bands are painted to indicate resistance and tolerance values. Using this approach, you might write the final sentence of the introduction as follows:

> The resistor consists of the following parts: two wire leads, the carbon element, the casing, and the color bands.

Or if you are using, for example, an inside-out spatial approach, the final sentence of the introduction might read this way:

> The resistor consists of the following parts: the carbon element, the wire leads, the casing, and the color bands.

Both approaches are fine as long as they are logical and make sense to the reader. In any case, once the parts have been listed, move on to the discussion section, where detailed descriptions of the parts are provided.

Discussion

The discussion section of a mechanism description precisely describes, in the necessary detail, each part of the mechanism. It always follows the organizational pattern established by the listing of parts at the end of the introduction. For example, if we are using inside-out spatial ordering, the discussion section will have four subsections: the carbon element, the leads, the casing, and the color bands.

The discussion section is laid out in a relatively simple manner in the order of the listing: Create a subsection for each part, and then describe the first part in the first subsection, and so forth.

Carbon Element

First, define the part with a technical definition:

> The *carbon element* is the capsule of resistive material that converts electrical energy to heat.

Second, provide an extension that describes the part's function and gives any needed theory:

> The carbon element serves as the primary active component of the resistor and provides the necessary 33,000 ohms of resistance. The element functions by blocking, to some degree, the flow of free electrons passing through it. The energy released by these blocked free electrons is then dissipated in the form of heat.

Third, describe the part's shape, color, material, finish, texture, and size, using precise measures and descriptors. Be sure to include necessary visuals such as diagrams and photographs. Two visuals will be added to this example later in this chapter; but for now, concentrate on describing the part's physical attributes:

> The carbon element is cylindrically shaped and is 2.4 centimeters long with a diameter of .31 centimeter. It is composed of finely ground carbon particles mixed with a ceramic binding compound. The element is gray with a dull, matte finish.

Fourth, transition to the next part by showing how this part relates to it:

> The carbon element is electrically connected to the leads.

Now handle the rest of the parts in similar fashion.

Leads

Define the part:

> The *leads* are two conductive wires connected to opposite ends of the carbon element.

Describe the function of the leads and provide any needed theory:

> The leads have two functions. First, they provide electrical connectivity from the carbon element to the circuit; and second, they provide a mechanical means of mounting and supporting the resistor in the circuit environment.

Describe the part in detail:

> The leads, which have a dull, silver color and smooth texture, are composed of tinned copper wire. Each lead is 4 centimeters long and is 20 gauge in thickness.

Transition to the next part:

> The leads connect to the carbon element through the ends of the casing.

Casing

Define the part:

> The *casing* is a cylindrical enclosure that surrounds the carbon element.

Describe its function and provide any needed theory:

> The function of the casing is twofold. First, it physically protects and insulates the carbon element from the outside environment. Second, it provides the heat-exchanging medium needed to dissipate the thermal energy generated by the carbon element.

Describe the part in detail:

> The casing is a brown, plastic cylinder that is 2.5 centimeters long, with a .312-centimeter inside diameter and a .52-centimeter outside diameter. It snugly fits over the carbon element.

And transition to the next part:

> The outside of the casing is circumscribed by four color bands.

Color Bands

Define the part:

> The *color bands* are visual indicators that describe the resistance and tolerance of the resistor.

Describe the function of the bands and provide the needed theory:

Using the standard color code for commercial, four-band resistors, and starting with the band at the edge of the cylinder, the first three bands represent the value of the resistor in ohms. The fourth band indicates the tolerance or accuracy of the resistor.

Describe the part in detail:

Each color band is .1 centimeter wide and is circumscribed on the outside of the casing and parallel to the edge of the casing. Each color band is smooth and shiny. The first color band, which starts flush with one end of the casing, is orange and represents a value of 3. The second color band, which starts .1 centimeter away from the inside edge of the first color band, is orange and also represents a value of 3. The third color band, which starts .1 centimeter away from the inside edge of the second color band, is orange and represents a multiplier of 1000. The fourth color band, which starts .1 centimeter away from the inside edge of the third color band, is gold and represents a tolerance of 5 percent.

Now that all the parts have been described, it is time to add a conclusion.

Conclusion

The final section of the mechanism description serves two purposes: It summarizes the description of the mechanism, and it provides a sense of finality to the document.

First, briefly summarize the mechanism's function and relist its parts:

The 33-kilohm, 1-watt carbon resistor is a circuit component that impedes the flow of electric current through the use of a carbon element. The resistor is made up of four parts: the carbon element, which impedes the flow of current by converting a portion of the electrical energy applied to heat; the wire leads, which electrically connect the element to the circuit and support the resistor mechanically; the casing, which encloses and insulates the element and dissipates heat from it; and the color bands, which indicate the resistance and tolerance of the device.

Now give a sense of finality to the paper. Include a sentence that by tone and content indicates to the reader that the mechanism description is complete:

> Together, these parts form one of the most commonly used circuit components in electronic systems today.

This final sentence tells the reader not to look for anything else because the document is ending; it is a courtesy to the reader.

Visuals and Mechanism Descriptions

When you are designing visuals such as diagrams, figures, and photographs in support of a mechanism description, be sure to include only information that directly relates to the mechanism description and is specifically keyed to the description provided in the text. Do not include visuals that do not match the mechanism description in subject matter or terminology. Also, avoid visuals that include too much or too little complexity for the level of discussion in the paper.

The following is the complete mechanism description. Note that two visuals have been added to enhance the description, a photograph and a diagram. Pay particular attention to these factors:

• These visuals specifically relate to the text discussion, and they are referred to in the text before actually being used. The reader should never encounter a visual and wonder why it is there or what it is for. Always tell your reader specifically when to look at a visual, and try to do so before the point in the paper where the visual actually appears.

- Each visual has an assigned sequence number and name, such as "Figure 4.1 Parts of the resistor." These labels reference precisely the visuals that are included. By the way, it is a good idea to use compound numbers—to show both the section or chapter number and the sequence number. You will notice in the following example that the first figure is labeled Figure 4.1. The 4 indicates that the visual occurs in the fourth chapter of this book, and the 1 refers to the sequence in which the figure occurs. Numbering visuals in this way allows you to add or delete visuals in one section without having to renumber all the visuals in subsequent sections. Also, if possible, run separate sets of sequence numbers for each type of visual. For example, figures should have their own set, as should tables and photographs.

- Each visual must be included for a purpose. In the following description, Figure 4.1 provides a visual overview of the device and labels its main parts, using a cutaway for the internal carbon element. Figure 4.2 provides an X-ray view of the resistor that shows the various physical dimensions.

| **Description of a 33-kilohm, 1-Watt Carbon Resistor** | **Putting It All Together** |

Introduction

The 33-kilohm, 1-watt carbon resistor is a circuit component that impedes the flow of electric current. The resistor impedes the flow of current by converting a portion of the electrical energy flowing through it to thermal energy, or heat. This particular resistor can safely convert 1 watt of electrical energy to heat.

The 33-kilohm, 1-watt carbon resistor looks like a small cylinder with wire leads extending from each end. The casing's surface is composed of smooth, brown plastic with a shiny finish. Four equally spaced color bands (three orange, one gold) circumscribe the cylinder, starting at one end.

The resistor consists of the following parts: the carbon element, the wire leads, the casing, and the color bands (see Figure 4.1).

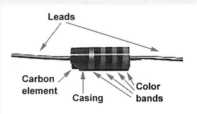

Figure 4.1
Parts of the resistor.

Discussion

For the following discussion, refer to Figure 4.2 for an X-ray view of the resistor's parts, along with its physical dimensions.

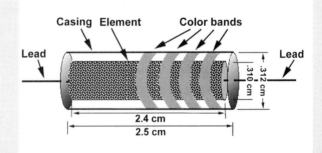

Figure 4.2
X-ray view of resistor.

Carbon Element

The carbon element is the capsule of resistive material that converts electrical energy to heat. The carbon element serves as the primary active component of the resistor by providing the necessary 33,000 ohms of resistance. The element functions by blocking, to some degree, the flow of free electrons passing through it. The energy released by these blocked free electrons is then dissipated in the form of heat.

The carbon element is cylindrically shaped and is 2.4 centimeters long with a diameter of .31 centimeter. It is composed of finely ground carbon particles mixed with a ceramic binding compound. The element is gray with a dull, matte finish. The carbon element is electrically connected to the leads.

Leads

The leads are two conductive wires connected to opposite ends of the carbon element. The leads have two functions. First, they provide electrical connectivity from the carbon element to the circuit; and second, they provide a mechanical means of mounting and supporting the resistor in the circuit environment. The leads, which have a dull, silver color and smooth texture, are composed of tinned copper wire. Each lead is 4 centimeters long and is 20 gauge in thickness. The leads connect to the carbon element through the ends of the casing.

Casing

The casing is a cylindrical enclosure that surrounds the carbon element. The function of the casing is twofold. First, it physically protects and electrically insulates the carbon element from the outside environment. Second, it provides the heat-exchanging medium needed to dissipate the thermal energy generated by the carbon element. The casing is a brown, plastic cylinder that is 2.5 centimeters long, with a .312-centimeter inside diameter and a .52-centimeter outside diameter. It snugly fits over the carbon element. The outside of the casing is circumscribed by four color bands.

Color Bands

The color bands are visual indicators that describe the resistance and tolerance of the resistor. Using the standard color code for commercial, four-band resistors and starting with the band at the edge of the cylinder, the first three bands represent the value of the resistor in ohms. The fourth band indicates the tolerance or accuracy of the resistor. Each color band is .1 centimeter wide and is circumscribed on the outside of the casing and parallel to the edge of the casing. Each color band is smooth and shiny.

The first color band, which starts flush with one end of the casing, is orange and represents a value of 3. The second color band, which starts .1 centimeter away from the inside edge of the first color band, is orange and also represents a value of 3. The third color band, which starts .1 centimeter away from the inside edge of the second color band, is orange and represents a multiplier of 1000. The fourth color band, which starts .1 centimeter away from the inside edge of the third color band, is gold and represents a tolerance of 5 percent.

Conclusion

The 33-kilohm, 1-watt carbon resistor is a circuit component that impedes the flow of electric current through the use of a carbon element. The resistor is made up of four parts: the carbon element, which impedes the flow of current by converting a portion of the electrical energy applied to heat; the wire leads, which electrically connect the element to the circuit and support the resistor mechanically; the casing, which encloses and insulates the element and dissipates heat from it; and the color bands, which indicate the resistance and tolerance of the device. Together, these parts form one of the most commonly used circuit components in electronic systems today.

Specifications and Functional Mechanism Descriptions

Functional mechanism descriptions focus primarily on those physical attributes that relate directly to the device's operating parameters and capabilities. These descriptions provide precise technical information for a mechanism, including standards for its use. Functional descriptions, which are used frequently to provide specifications, normally include the following: a general description of the mechanism, visuals related to the device's structure and function, and a listing of physical attributes that directly affect or describe the mechanism's operational profile, capabilities, and limitations.

The following example is a codified form of a functional mechanism description called a *specification sheet*. This example provides a one-page functional mechanism description for the 16XL1000000 transmitting tube. It is designed as a quick reference for an expert audience; as such, it assumes substantial knowledge about high-power transmitting tubes on the part of the reader.

Tube 16XL1000000 Specification Sheet

General Description

The 16XL1000000 high-power, ceramic tetrode is a class C transmitting tube for broadcast service at frequencies up to 150 megahertz. The tube has a thoriated-tungsten mesh filament mounted on water-vapor-cooled supports. The outer casing is composed of steel and ceramic and is circumscribed with heat radiation fins around the base of the anode dome structure extending up from the plate (P) electrode. Maximum anode dissipation is 1000 kilowatts in continuous service with high-level amplitude modulation. Water-vapor cooling ports extend from the base. Flange electrodes are used for plate, screen grid, and ground; coaxial terminals provide connectivity for the control grid and RF filament. See Figure 4.1a and b.

Figure 4.3 - 16XL1000000 Composite.

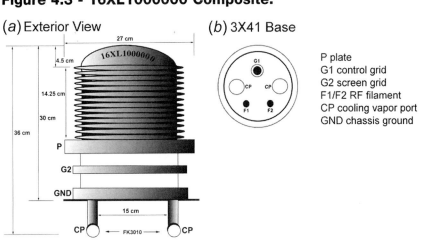

(a) Exterior View   (b) 3X41 Base

P plate
G1 control grid
G2 screen grid
F1/F2 RF filament
CP cooling vapor port
GND chassis ground

Specifications

Maximum useful frequency = 300 megahertz

Frequency for ratings (max.) = 200 megahertz

Plate voltage (typ.) = 22 kv

Amplification factor (G1−G2) = 6.2

Plate dissipation (max.) = 1,200,000 watts

G2 dissipation (max.) = 10,000 watts

Plate current (typ.) = 60 amperes

G1 dissipation (max.) = 2000 watts
Heat dissipation (max.) = 800,000
    watts
Cooling = water vapor
Cooling port connector = FK-3010
    Core temperature (max.) = 2008C
Weight = 98 kilograms
Capacitance (cathode GND) in = 1.1
    millifarad
Capacitance (grid GND) out = 175
    picofarad
Capacitance (grid GND) in = 490
    picofarad

Capacitance (feed-through) = 6
    picofarad
Capacitance (cathode GND) out = 185
    picofarad
Capacitance (feed-through) = 0.8
    picofarad
Filament = thoriated-tungsten mesh
Filament power = 16.2 volts @ 600
    amperes
Base = 3X41 coaxial
Filament RF connector = FK-3000
Grid RF connector = FK-3010

- Have I defined the mechanism, and have I extended this definition with any theory or principles necessary for my reader's understanding?
- Have I included theory or principles that either are not needed or are above or below the level of my reader?
- Have I described the mechanism's overall function or purpose?
- Have I described the overall appearance of the mechanism in terms of its shape, color, material, finish, texture, and size?
- Have I listed the main parts in the introduction?
- Have I defined each of these parts?
- Have I described each part's function or purpose?
- Have I discussed the needed theory or operating principles for each part?
- Have I precisely described each part's physical structure?
- Have I provided transitions from one part to the next?
- Have I concluded by summarizing the mechanism's function and parts?
- Have I given a sense of finality to the description?

**Mechanism
Description
Checklist**

Exercise Apply the checklist just given to the following mechanism description of the FinkelBOAT 688 Attack Submarine. What problems exist? Are any parts missing? Do the definitions work? Are any of them circular? Do some require additional definitions or extensions? Can you determine the intended audience and the purpose of the description? Pay particular attention to not only the size and complexity of this mechanism, but also the purpose, presentation, and quality aspects of the description. What about the visual? Is the level of complexity appropriate for this description? Is the visual properly marked? Does it appear in the correct place? What purpose does it serve? How well integrated is the visual into the text's description? Also, what sections provide precise technical writing, and what sections feature abstract, imprecise concepts?

Mechanism Description of a FinkelBOAT 688 Attack Submarine

Introduction

The FinkelBOAT 688 submarine is a nuclear-powered, attack, undersea watercraft characterized by high-speed, ultraquiet operation, high cost, and an awesome array of conventional and special weapons. The submarine, which is used primarily for peaceful undersea research, has a top submerged speed of 44 knots and carries both UGM-109 cruise missiles and Mk-48 torpedoes. The FinkelBOAT 688 consists of a main tube, two hemispheric caps, a propeller, and a fairwater.

Discussion
Main Tube

The main tube is a claustrophobic cylinder of pure YM-31 high-tensile titanium that provides the body of the boat. The main tube is 75 feet long, 6 inches thick, and has an inside diameter of 33 feet. The main tube is coated with an anechoic/decoupling coating designed to defeat active sonars and contain internal noise. At each end of the main tube are hemispheric caps (Figure 4.4).

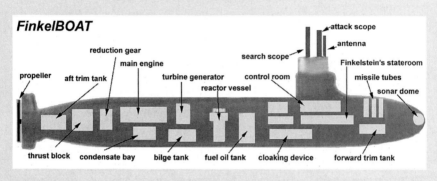

Figure 4.4
FinkelBOAT.

Hemispheric Caps

The hemispheric caps are cones of pure YM-31 high-tensile titanium that enclose both the stern and bow ends of the main tube, thereby completing the substructure of the boat. The caps are tapered and welded to the main tube's barrel sections. The boat's propeller is located directly behind the stern hemispheric cap.

Propeller

The propeller is a seven-blade propeller that provides high thrust with little or no cavitation. The propeller, which is made of a special bronze alloy, is 30 feet in diameter and can silently move 500 acre-feet of water per second at flank speed. On the top of the substructure, and 55 feet forward toward the bow hemispheric cap, is the fairwater.

Fairwater

The *fairwater* is the raised, enclosed observation post of the submarine that supports periscopes and communications antennas and that also provides a modest bridge area. Often called a sail or conning tower, the fairwater, which is 25 feet high, 15 feet long, and 10 feet wide, is also composed of pure YM-31 high-tensile titanium.

Conclusion

As Figure 4.4 shows, the FinkelBOAT has lots of parts that work together to make a submarine.

05

Description of a Process

Process descriptions are similar to mechanism descriptions in organizational structure, but they differ in content. Whereas mechanism descriptions detail the physical attributes of a mechanism, process descriptions describe the steps of a process.

Processes are important in technical writing because much of what needs to be communicated involves unfolding events or actions. In fact, process descriptions are key elements of many technical reports. Process descriptions can include objective, third-person portrayals of events that do not directly involve the reader, such as explaining to your teenager how a modern car operates. They can also be second-person descriptions for human involvement that provide specific instructions so that the reader can perform or accomplish the process, such as showing your teenager how to drive the car. Normally, giving specific instructions is a more challenging task.

What Is a Process Description?

Process descriptions are precise portrayals of events occurring over time that lead to some outcome. Process descriptions describe either the steps in the operation of a mechanism or the steps of a conceptual process. Unlike mechanism descriptions, process descriptions of a mechanism in operation focus less on the physical attributes of the mechanism, and more on a mechanism's function and how the parts work together. For example, a description of a car's four-cycle, internal combustion engine in operation would not focus on the parts (such as the pistons, rods, cylinders, valves, and spark plugs); rather, the description would focus on the steps of the engine's operation (such as intake stroke, compression stroke, ignition stroke, and exhaust stroke). By the same token, a description of how to drive the car might include steps such as fastening the seat belt, starting the engine, engaging the foot brake, releasing the parking brake, and putting the transmission in gear.

As with mechanisms in operation, conceptual process descriptions discuss the steps of a process, but these steps do not involve a physical mechanism. For example, a process description for operations research might well include a decision tree; however, this tree is not the kind you have to plant, water, and fertilize. In fact, it is not a thing at all. It exists only conceptually in the imagination.

Whether a process involves something physical is irrelevant. The organization of a process description is basically the same, whether it describes a mechanism in operation or a conceptual process. Additionally, the general organization of the process description is the same whether it provides a third-person account of what happens in the process or second-person instructions on how to accomplish the process.

Outline of a Process Description

Notice that the two outlines for process descriptions are similar. Outline 5.1 lays out the pattern for describing the operation of a mechanism, such as an air conditioning system that includes compression, condensation, expansion, and evaporation as its steps. Outline 5.2 provides a similar pattern for describing a conceptual process, such as a sort algorithm that includes the iterative steps of evaluation, identification, and insertion.

Outline 5.1 Process Description of a Mechanism in Operation

Introduction
- Define the mechanism with a technical definition (see Chapter 3) and add extensions to discuss any theory or principles necessary for the reader to understand what you are saying. Make sure you include only what the reader needs for your purpose.
- Describe the purpose, function, and operation of the mechanism.
- List the major steps of the mechanism's operation.

Discussion
- Step 1
  - Define the step with a logical definition, and discuss it in reasonable depth.
  - Describe the equipment, material, or concepts involved in this step.
  - Describe what happens during this step.
  - Show the relationship between this step and the next step with a transition statement.
- Steps 2–n
  - For each remaining step, repeat the pattern established for step 1. The final step discussed may not have a transition, unless it cycles back to the first or subsequent step.

Conclusion
- Briefly summarize the mechanism's function and the major steps of its operation.
- Give a sense of finality to the paper.

Outline 5.2 Description of a Conceptual Process

Introduction
- Define the process with a technical definition (see Chapter 3), and add extensions to discuss any theory or principles necessary for the reader to understand what you are saying. Make sure you include only what the reader needs for your purpose.
- Describe the purpose and function of the process.
- List the major steps of the process.

Discussion
- Step 1
  - Define the step with a logical definition, and discuss it in reasonable depth.
  - Describe the purpose and function of this step.
  - Describe what happens during this step.
  - Show the relationship between this step and the next step with a transition statement.
- Steps 2–*n*
  - For each remaining step, repeat the pattern established for step 1. The final step discussed may not have a transition, unless it cycles back to the first or subsequent step.
Conclusion
- Briefly summarize the function and the major steps of the process.
- Give a sense of finality to the paper.

The following three descriptions illustrate the use of these outlines to describe processes. The first example describes the operation of a mechanism, in this case an air conditioner. The second example describes the purely conceptual process of a sort algorithm, which does not involve a material mechanism. The third example (the Threaded Example) provides a short description of the 16XL1000000 transmitting tube's operation.

Description of a Mechanism in Operation

To provide an example for using Outline 5.1, this chapter will describe the process of the operation of a mechanism, in this case a notional air conditioning system. We will add visuals later, once the process has been developed. The audience for this description is a general, technical reader who does not have specific expertise in air conditioning, fluid mechanics, or thermodynamics.

Introduction

Begin by introducing the mechanism with a logical definition; then add extensions to provide needed theory or principles:

An *air conditioner* is a mechanical device used to refrigerate a controlled environment. The air conditioner accomplishes this refrigeration by transferring heat within the environment to an area outside the environment.

Next describe the function and operation of the mechanism:

The air conditioner transfers heat by using a fluid refrigerant. This refrigerant is pumped through both the controlled environment and the outside area. At the same time, the refrigerant is cycled at strategic points between liquid and vaporous states. This change in state provides the means for transferring thermal energy.

You may want to list the primary operational parts of the mechanism, especially if they relate directly to the process steps. In any case, be sure to list the major steps of the mechanism's operation in the order in which they will be described. Choosing this order may not be easy. Ideally, the steps will follow a logical time line from the start to the finish of the process; however, in some cases, such as transactional, iterative, and branching processes, that may not be possible. In such cases you will have to describe the process steps in the way that is clearest for your reader and the purpose at hand.

The air conditioner's operation is centered on four major components, including the compressor, condenser coil, expansion valve, and evaporator coil. The operation of the air conditioner relates directly to these parts and includes the following steps: compression, condensation, expansion, and evaporation.

Discussion

In the discussion section, treat each step as a separate subsection. Define each step of the process; then extend these definitions as necessary to address the equipment, material, or concepts involved. Be sure to tell your reader clearly what happens in each step, and provide linking transitions between the steps.

Compression

Discuss compression in reasonable depth for the audience and purpose at hand. First define compression, and then deal with the equipment and concepts involved. Describe what happens during compression, and then show the relationship with the next step, condensation:

> *Compression* is a fluid-dynamics process in which a given volume of refrigerant vapor is forced to occupy a smaller volume of space. Compression occurs when the compressor forces hot refrigerant vapor under pressure into the compression chamber. The chamber is composed of a cylinder/valve arrangement. The piston draws refrigerant into the cylinder through an intake valve; the intake valve closes; and the piston pushes up into the cylinder, compressing the refrigerant vapor. The vapor then exits through an exhaust valve and enters the condenser coil, where the condensation step occurs.

Condensation

Describe condensation in much the same way as you dealt with compression. Start with a technical definition of condensation, and then extend this definition as necessary to describe the parts of the mechanism involved and the concepts and theory the reader may need. Next describe what happens in this step of the process, and transition to the third step, expansion:

> *Condensation* is a fluid-dynamics process in which the hot refrigerant vapors, pumped from the compressor to the condenser coil, cool and change to a liquid. As this change occurs, the heat of condensation is released from the refrigerant in the

condenser coil, heating the coil. This heat, in turn, is drawn from the coil by a fan, which passes relatively cooler air across the coil, picking up the heat and venting it outside. The refrigerant liquid then flows through a closed loop to the expansion valve, where rapid expansion occurs.

Expansion

Treat this third step in the same manner as the first two. First define expansion, and then extend the definition to describe the parts of the mechanism involved and to provide needed theory and concepts. Explain what happens during this step, and provide a transition to the fourth step, evaporation:

> *Expansion* is a fluid-dynamics process in which the condensed liquid refrigerant, under relatively high pressure from the compressor, is forced through an expansion valve into an area of substantially lower pressure. The expansion valve acts as a nozzle, constricting and then accelerating the liquid refrigerant until it passes through a threshold where the constriction is removed and rapid expansion occurs. At this point the expanding refrigerant enters the evaporator coil, where it will cool and change state again.

Evaporation

Finally, define the last step, evaporation; extend the definition; and describe what happens, as in earlier steps. However, because this process is cyclical and repetitive, the transition from this last step should link back to the first step.

> *Evaporation* is a fluid-dynamics process in which the rapidly expanding refrigerant liquid changes to a vapor. As the liquid enters the evaporator coil, it also enters an area of substantially lower pressure. As a result, it vaporizes and in the process absorbs heat. Air in the controlled environment is circulated across this coil, which, in turn, absorbs heat from the air. The fan distributes the resulting cooler air

throughout the controlled environment. The refrigerant vapor in the evaporator coil is then drawn back through the closed loop to the compressor, at which point the entire cycle repeats.

Conclusion

The last part of the process description summarizes the process and provides a sense of finality as a courtesy to the reader:

> The operation of an air conditioner involves four steps. First, the compressor pumps refrigerant under pressure into the condenser coil. Here it is liquefied, giving up heat that is removed by a fan circulating air over the condenser coil. The liquid refrigerant moves through a closed loop, through an expansion valve, and into the lower-pressure evaporator coil. Here the refrigerant changes to a vapor, absorbing heat from air passing over the evaporator coil. This air then cools the controlled environment, while the refrigerant is drawn back into the compressor. At this point the cycle is complete, and the process repeats.

Visuals and Process Descriptions

Designing visuals for process descriptions follows the same rules and has the same considerations as designing visuals for mechanism descriptions—except for one significant difference. Process visuals usually need to show something happening and, as such, tend not to be static representations of something seen in a "slice of time." Often you can show movement through space and time with a technique as simple as a well-placed arrow. At other times you may need a coherent series of visuals to demonstrate your point. Additionally, well-integrated captions within the visual can help explain it.

**Putting It All
Together**

Description of the Operation
of an Air Conditioner

Introduction

An *air conditioner* is a mechanical device used to refrigerate a controlled environment. The air conditioner accomplishes this refrigeration by transferring heat within the environment to an area outside the environment. The air conditioner transfers heat, using a fluid refrigerant. This refrigerant is pumped through both the controlled environment and the outside area. At the same time, the refrigerant is cycled at strategic points between liquid and vaporous states. This change in state provides the means for transferring thermal energy.

The air conditioner's operation is centered on four major components, including the compressor, condenser coil, expansion valve, and evaporator coil. The operation of the air conditioner relates directly to these parts and includes the following steps: compression, condensation, expansion, and evaporation. See Figure 5.1.

Discussion
Compression

Compression is a fluid-dynamics process in which a given volume of refrigerant vapor is forced to occupy a smaller volume of space. Compression occurs when the compressor forces hot refrigerant vapor under pressure into the compression chamber. The chamber is composed of a cylinder/valve arrangement. The piston draws refrigerant into the cylinder through an intake valve; the intake valve closes; and the piston pushes up into the cylinder, compressing the refrigerant vapor. The vapor then exits through an exhaust valve and enters the condenser coil, where the condensation step occurs.

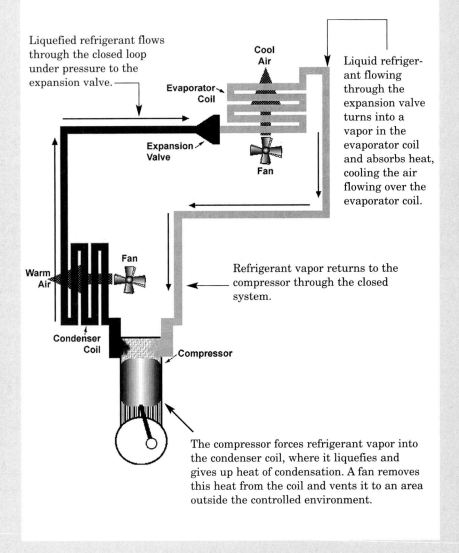

Liquefied refrigerant flows through the closed loop under pressure to the expansion valve.

Cool Air

Evaporator Coil

Liquid refrigerant flowing through the expansion valve turns into a vapor in the evaporator coil and absorbs heat, cooling the air flowing over the evaporator coil.

Expansion Valve

Fan

Warm Air

Fan

Refrigerant vapor returns to the compressor through the closed system.

Condenser Coil

Compressor

The compressor forces refrigerant vapor into the condenser coil, where it liquefies and gives up heat of condensation. A fan removes this heat from the coil and vents it to an area outside the controlled environment.

Figure 5.1
Air conditioner. Note the use of text blocks to provide captions that key the steps of the process to the diagram.

Condensation

Condensation is a fluid-dynamics process in which the hot refrigerant vapors, pumped from the compressor to the condenser coil, cool and change to a liquid. As this change occurs, the heat of condensation is released from the refrigerant in the condenser coil, heating the coil. This heat, in turn, is drawn from the coil by a fan, which passes relatively cooler air across the coil, picking up the heat and venting it outside. The refrigerant liquid then flows through a closed loop to the expansion valve, where rapid expansion occurs.

Expansion

Expansion is a fluid-dynamics process in which the condensed liquid refrigerant, under relatively high pressure from the compressor, is forced through an expansion valve into an area of substantially lower pressure. The expansion valve acts as a nozzle, constricting and then accelerating the liquid refrigerant until it passes through a threshold where the constriction is removed and rapid expansion occurs. At this point, the expanding refrigerant enters the evaporator coil, where it will cool and change state again.

Evaporation

Evaporation is a fluid-dynamics process in which the rapidly expanding refrigerant liquid changes to a vapor. As the liquid enters the evaporator coil, it also enters an area of substantially lower pressure. As a result, it vaporizes and in the process absorbs heat. Air in the controlled environment is circulated across this coil, which, in turn, absorbs heat from the air. The fan distributes the resulting cooler air throughout the controlled environment. The refrigerant vapor in the evaporator coil is then drawn back through the closed loop to the compressor, at which point the entire cycle repeats.

Conclusion

The operation of an air conditioner involves four steps. First, the compressor pumps refrigerant under pressure into the condenser coil. Here it is liquefied, giving up heat that is removed by a fan circulating air over the condenser coil. The liquid refrigerant moves through a closed loop, through an expansion valve, and into the lower-pressure evaporator coil. Here, the refrigerant changes to a vapor, absorbing heat from air passing over the evaporator coil. This air then cools the controlled environment, while the refrigerant is drawn back into the compressor. At this point the cycle is complete, and the process repeats.

Description of a Conceptual Process

The following is a brief, notional example of an iterative, conceptual process. Computer programmers use this particular process, which is roughly analogous to a selection sort algorithm, to put lists of unsorted data into either ascending or descending order. In reality, the actual implementation of a selection sort algorithm in modern, higher-level languages is far more complex and efficient than what is represented here.

This example, which integrates two visuals into the description, is written for a general, technical audience who does not have specific expertise in computer programming and information systems. For illustration, this example process for sorting data in descending order uses the following list of six numbers:

6, 32, 8, 19, 3, 20

<table>
<tr><td>

Process Description of Selection Sort

</td><td>

Putting It All Together

</td></tr>
</table>

Introduction

Selection sort is an iterative process in which a list of unsorted data is placed in either ascending or descending order. Selection sort gets its name from the way in which the data are sorted. During each iteration, the smallest value from a list of unsorted numbers is selected and then inserted in the front of the list. As the process continues, the sorted area in the front of the list gets larger, while the remaining unsorted area gets smaller. At some point, when all the numbers from the unsorted list have been selected and inserted, the list is sorted, and the process is complete. The selection sort process involves the following iterative steps: evaluation, identification, and insertion. See Figure 5.2 for a flowchart of the process.

Discussion

The following example sorts the list in descending order.

Evaluation

Evaluation is the scanning process in which the list of numbers is tested for proper order. This step normally involves a pass through the list where a series of comparisons of adjacent numbers determines whether the list is ordered. If the list is ordered, the process is complete and ends. If the list is not ordered, then the lowest value is identified.

Identification

Identification is the scanning process by which the lowest value in the list is located. Normally, the lowest value is identified by a series of comparisons and swaps. Once identified, this value can be inserted.

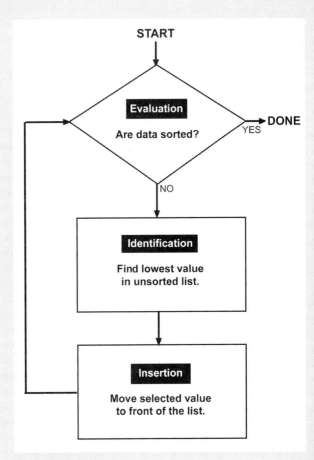

Figure 5.2
Selection sort flowchart.

Insertion

Insertion is a relocation process in which the lowest value in the unsorted area of the list is placed or inserted in the front of the list. On the first iteration, this value starts the ordered or sorted area of the list, while the remaining numbers constitute the unordered or unsorted area of the list. Once the lowest value has been inserted, the process returns to the evaluation step, where the iteration continues until no unordered numbers remain (see Figure 5.3 for an example).

Conclusion

Selection sort is an iterative process in which a list of unsorted data is placed in either ascending or descending order. During each iteration, the list is evaluated to see if it is, in fact, ordered. If not, the smallest value from a list of unsorted numbers is selected and then moved to the front of the list. As the process continues, the sorted area in the front of the list gets larger, while the remaining unsorted area gets smaller. At some point, when all the numbers from the unsorted list have been moved, the list is sorted and the process is complete. Insertion sort is one of the many algorithms used by computer programmers to sort lists of data.

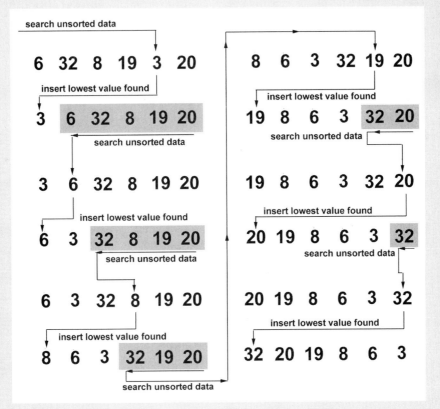

Figure 5.3
Selection sort.

Operation of the 16XL1000000 Megatube

The following process description of the 16XL1000000's operation is written for a technical audience. The assumption is that the reader will understand the meaning and concepts associated with terms such as *RF, molybdenum, thoriated tungsten, electrostatics, megawatt,* and *resonance.*

Figure 5.4 - 16XL1000000 Schematic.

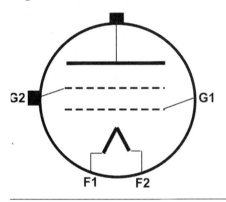

Introduction

The *16XL1000000 Megatube* is a ceramic and metal, high-power, transmitting tetrode that can produce very high levels of RF at frequencies up to 300 megahertz. The tube uses a thoriated-tungsten mesh filament (F) in a steel and ceramic outer casing. In addition to the filament, the active elements include a molybdenum wire control grid (G1) that regulates the flow of electrons; a molybdenum wire screen grid (G2) that isolates the input from the output and reduces unwanted feedback; and a graphite plate (P) that acts as the positively charged anode, attracting and collecting electrons flowing from the filament through the grid structures. The tube functions as a power amplifier by controlling a large flow of electrons (22 kilovolts at 60 amperes) through P with a relatively low-level input signal at G1; in fact, it can generate an output of 1.24 megawatts with an input signal of 2 kilowatts. The operation, which is centered on the active elements (F, G1, and P), includes the following steps: emission, control, and collection. (See Figures 5.4 and 5.5.) (*Note:* Because G2's function is only to provide isolation screening, it is not included in the following discussion.)

Figure 5.5 - 16XL1000000 Diagram

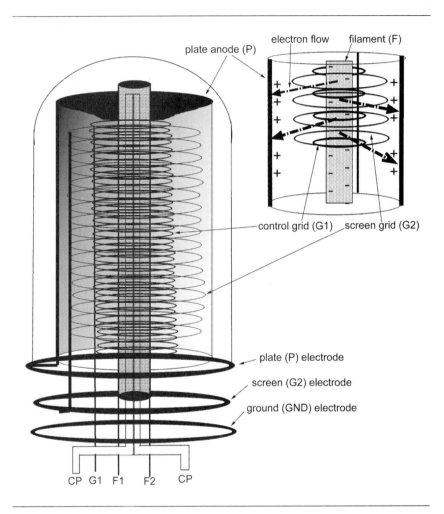

Discussion

EMISSION

Emission is a thermal process during which an electron stream flows away from F. F is composed of coiled thoriated-tungsten wire. When 16.2 volts is applied across F, a current of 600 amperes flows, causing F to heat to about 2400°C. The thorium in F migrates to F's outer surfaces, where it emits a "cloud" of electrons. P then draws these electrons through G1, allowing G1 to control the electron flow.

CONTROL

Control is the electrostatic process by which the input signal applied to G1 regulates the flow of electrons emitted from the filament. The electron cloud formed around F flows away from F, passing through G1 on its way to P. The input signal applied to G1 causes G1's polarity to shift relative to P, thereby allowing the input signal to control the number of electrons flowing to the plate. As G1 is made more negative, P is less able to collect electrons and its current decreases. When G1 is made less negative, P's current increases because P is better able to collect electrons.

COLLECTION

Collection is an electrostatic process by which P attracts and accumulates electrons flowing from F through G1. When 22 kilovolts is applied between P and F, and P's output circuitry is tuned to resonance, P collects enough electrons to establish an average current flow of 60 amperes. This plate current represents the amplified output of the tube.

Conclusion

The operation of the 16XL1000000 high-power transmitting tube is centered on three steps: emission, control, and collection. Current passes through F, causing it to heat up and emit electrons. The electrons flow away from F through G1 and are then collected by P. Varying polarities of the input signal applied to G1 cause variations in P's ability to attract and collect electrons. The resulting variation in current flowing through P represents the amplified output of the tube.

- Have I defined the mechanism or the process, and have I extended this definition with any theory or principles necessary for the reader's understanding? Have I included theory or principles that either are not needed or are above or below the level of the reader?
- Have I described the process's overall function or purpose?
- Have I described the overall function or operation involved in the process?

Process Description Checklist

- Have I listed the main steps of the process in the introduction?
- Have I defined each of these steps?
- Have I described each step's function or purpose?
- Have I discussed the needed theory or operating principles for each step?
- Have I precisely described what happens in each step?
- Have I provided transitions from one step to the next step?
- For iterative processes, have I clearly shown when branching occurs and when the process is complete?
- Have I concluded by summarizing the process's purpose and function?
- Have I given a sense of finality to the description?

Exercise Apply the checklist just given to the following process description of the QuadFINKEL figure skating jump. (Of course, the author quite vainly named the jump after himself. If Ulrich Salchow, Alois Lutz, and Axel Paulson can do it, so can he—except, of course, he does not know how to skate.) In particular, look at the following:

- Is the subject a process? If so, what areas of this process description are not technical writing? Do any areas deal with abstract, imprecise value judgments and concepts?
- For what audience was this description written? Is there a consistent audience? What areas of expertise would be required for the reader to understand the various parts? At what level of expertise is this paper written?

- How is the paper organized? Does the introduction properly introduce the process? Are
the steps well defined? Are they adequately
described? Does each step transition well to the
next step?
- Look at the visuals. One exists for each step of
the process. Are these visuals useful? Are they
properly marked? Do they have adequate label-
ing? Are they placed effectively in the document? Are they well integrated into the text's
description?

Process Description of a QuadFINKEL

Introduction

In figure skating, the *QuadFINKEL* is a competitive jump that combines a Lutz glide, Axel vault,
high-speed spin, and Salchow landing. The
QuadFINKEL is regarded by the skating world as
the most difficult maneuver in the history of the
sport. The QuadFINKEL, which requires a skater
of great power, skill, and courage, has six steps:
the glide, the turn, the vault, the spin, the stall,
and the landing (see Figure 5.6).

Discussion
The Glide

The *glide* is an acceleration maneuver by which
the skater establishes high-speed movement on the
right outside back edge. As the glide speed
approaches 55 miles per hour, the left hand is
inserted into the left pocket to reduce aerodynamic
drag, while the right hand is extended to provide
rudder control; see Figure 5.6a. When backward
velocity reaches 60 miles per hour, the skater is
ready to turn.

(a) glide   *(b)* turn   *(c)* vault

(d) spin   *(e)* stall   *(f)* landing

Figure 5.6
QuadFINKEL.

The Turn

The *turn* is a transition maneuver in which the skater's body rotates 180° counterclockwise while maintaining the original vector. Additionally, the skater shifts from the right back outside edge to the left front outside edge and spreads both arms in preparation for the vault; see Figure 5.6*b*.

The Vault

The *vault* is a propulsion maneuver in which the skater leaps off the right front outside edge, accomplishing an Axel-entry blade jump. Once airborne, the skater extends both hands up and away from the body to provide rotational stability and artistic elegance; see Figure 5.6c. This action prepares the skater for the spin.

The Spin

The *spin* is a rotational maneuver in which the airborne skater rapidly moves his or her extended arms and hands to create both substantial artistic impression and a low-pressure vortex with resulting counterclockwise torque; see Figure 5.6d. Generally, spin speeds exceeding 2000 revolutions per minute are possible using this technique. The skater then turns his or her arms to simulate a Bernoulli surface, which provides sufficient aerodynamic lift to sustain autogyrational flight. (*Note:* In the southern hemisphere the rotation would be clockwise, so the QuadFINKEL is not possible south of the equator.) Once the rotational torque is established and the skater has made $3\frac{1}{2}$ revolutions, he or she is ready for the stall.

The Stall

The *stall* is an aerodynamic maneuver in which the skater uses his or her outstretched hands as spoilers to slow rotation to the point where lift is no longer sufficient to sustain flight; see Figure 5.6e. This action is taken after $3\frac{1}{2}$ revolutions, so that a stall condition occurs at precisely $4\frac{1}{2}$ revolutions. The stall leads immediately to landing.

The Landing

Landing is the final jump maneuver by which the skater touches down on the right back outside edge; see Figure 5.6f. The skater's left hand is reinserted in the pocket while the outstretched right hand is used again for rudder control. The left edge must not touch the ice until the skater's speed has been reduced to zero. At this point, the QuadFINKEL jump is complete.

Conclusion

The QuadFINKEL is the most difficult of all competitive figure skating jumps. To do a Quad-FINKEL, the skater does an Axel vault off the left front outside edge after a Lutz acceleration glide on the right back outside edge. The jump uses sophisticated aerodynamic forces to sustain a high-speed spin, followed by a Salchow landing on the right back outside edge. Because of its hypnotic artistic impression and demanding technical qualities, it has been said that "one who achieves the QuadFINKEL always lands on the gold."

06

Proposals

Proposals are among the most important documents one can write. Persons and organizations that write effective proposals win grants, contracts, and jobs; persons and organizations that do not write effective proposals often just wind up "going away"—sometimes "far away." Proposals are important because they, directly or indirectly, provide the income that keeps us warm, dry, and well fed!

Proposals are specialized, technical business documents that offer persuasive solutions to problems. All technical documents are designed to communicate ideas objectively and clearly. Your goal in writing a proposal, however, goes beyond just precise communication. A proposal also needs to sell the reader on some idea—usually that he or she (or his or her organization) needs specific goods or services that you (or your organization) can provide.

To be successful, you normally need to do at least three things in any proposal you write:

What Is a Proposal?

1. Describe, identify, or refer to a problem that needs to be solved. The reader probably already knows that he or she has a problem. You still need to describe it, however, because the reader may not fully understand or appreciate the scope, magnitude, or complexity of the issues at hand. Additionally, by providing this description, you establish credibility by showing your reader that you understand the problem.

2. Offer a viable solution for the problem. You have to demonstrate to the reader that your proposed approach will, in fact, solve the problem effectively and efficiently.

3. Show that you can effectively implement this solution. The fact that you have an effective solution does not mean much if you cannot implement it. That is why, in any proposal, you must show that you have the skills and resources required to do what you are proposing. One of the best ways of demonstrating your capability is with successful prior performance. In other words, if you are proposing to design a Web page for your company, it would be a definite plus if you could point to other successful Web pages that you have designed. Showing that you have done the kind of thing you are proposing to do—and have done it well—can be persuasive.

Formal and Informal Proposals

Proposals are generally categorized as *formal* or *informal*. Formal proposals are normally large, comprehensive documents produced by a team of experts on behalf of an organization. Informal proposals are generally short documents of limited scope written by an individual.

Formal Proposals

The topic of formal proposals is well beyond the scope of this book. But what if your boss just told you to put together a formal proposal and you have no clue what to do? Well, quite honestly, your boss is a fool, and you are using the wrong book! Formal proposals are lengthy projects undertaken by skilled proposal writers functioning together in a well-structured, proposal team environment. These documents are prepared in response to a formal *request for proposal* (RFP). Proposal teams take great care to respond to each RFP requirement.

(Chapter 19 provides a more thorough discussion of team writing in this context.) Formal proposals can take many forms, but a typical one might include the following:

- An *executive summary* that synopsizes the substance of the proposal. Executive summaries often are written by senior decision makers such as a company's vice president. This executive summary is the only part of the proposal that many readers look at, so it can be very important. Executive summaries are discussed in greater detail in Chapter 12.
- A *technical volume* that lays out the proposed solution in detail. This section is normally written by a team of engineers and scientists who are responsible for solving the problem.
- A *management volume* that describes the organizational structure and key players who will implement the proposal, if accepted. This section is generally written by a team of experts in management theory and organizational structure.
- A *cost volume* that provides detailed analysis and data regarding the cost of implementing the proposed solution. This section is often written by a team of financial planners, auditors, comptrollers, and accountants.
- A *resources volume* that provides detailed analysis and data regarding both the human and the physical resources required to implement the proposed solution. This section is written by human resource experts, who are responsible for hiring people with the necessary skills, and by facilities experts, who locate, modify, or build the needed facilities. In some cases this section is included in the management volume.

Formal proposals frequently take months to produce at a cost of tens of thousands of dollars. Writing these proposals is a difficult task, especially because these documents are evaluated in an extremely competitive environment. Winning or losing is based on the quality of the proposed solution, the credibility of the proposing organization, and the estimated cost of implementing the proposal. In the end, the proposal representing the best overall value normally wins the contract.

Government Proposals

Perhaps the best example of just how challenging proposal writing can be involves writing proposals for the U.S. government, where federal tax dollars are involved. Writing proposals for the government is truly an exacting, unforgiving activity. The content, layout, and procedures are precisely spelled out, and evaluation of proposals is rigorous and comprehensive. Source selection panels composed of technical, management, human resource, financial, and other experts evaluate each proposal. They aggressively seek out all discrepancies between tasks and skills, skills and costs, management philosophy and organizational structure, and so on. They score each proposal based on the effectiveness of the solution, the risks involved, the costs in both time and dollars, and the capabilities and past performance of the proposing company. The goal of the source selection process is to select the proposal that offers the best value to the government.

Interaction between proposal writers and proposal evaluators (that is, between the company and the government) is strictly controlled. Even the appearance of impropriety can have serious legal implications. After preliminary interactions to evaluate the objectives of the project, the government issues an RFP that precisely spells out the problem to be solved and the parameters and constraints for any proposed solution. In some cases, proposals are presented as technical briefings

(Chapter 17), or they may be submitted electronically (Chapter 16). Larger projects may require multivolume documents.

In most cases, government proposals have specific deadlines and delivery requirements and must precisely meet all format and content specifications.

Informal Proposals

The informal proposal is the primary focus of this chapter. Written by individuals, not teams, informal proposals typically address a limited problem for which a relatively straightforward solution exists. Frequently, informal proposals take the form of a long letter or a short document.

Informal proposals may also be either *solicited* or *unsolicited*. A solicited proposal responds to a specific, often written, request. A homeowner might ask a contractor for a proposal to replace the roof on a house, or a company might ask a lighting firm for a proposal to illuminate a manufacturing area. With a solicited proposal, the problem has already been identified, and the decision to solve the problem has already been made.

Unsolicited proposals, on the other hand, are proposals that no one has asked for and, perhaps, that no one wants. The recipient has not decided to solve a problem and may not even realize that a problem exists—or may not want to realize it. Unsolicited proposals often come from within an organization; for example, an assembly line worker may send the supervisor a letter suggesting a change to a manufacturing process. Unsolicited proposals, as you might imagine, have less chance of being accepted.

Informal proposals can take many forms and can be organized in many different ways. The best advice is to carefully think through all aspects of the problem and your proposed solution, and to use your common sense. Also, follow any guidelines from your boss or an agency requesting the proposal.

If you do not have specific guidance, then Outline 6.1 provides a fairly typical approach you can use to organize informal proposals. We will look at each element of this outline and see how you might go about putting together such a proposal.

As a hypothetical example, imagine that the International Olympic Committee (IOC) has asked your organization, ExtremaLab, to provide computer modeling and analysis of the Quad-FINKEL figure skating jump (the Chapter 5 exercise). To respond to the IOC, your ExtremaLab division must create an informal (that is, internal) proposal to upgrade its computer systems. The following proposal, written for the experts at ExtremaLab, is geared to an audience knowledgeable in modeling and analysis, in the technology used to accomplish such modeling and analysis, and in the overall resources available to ExtremaLab.

Outline 6.1 Informal Proposals

Introduction
- Purpose   Describe the reason for writing this report.
- Background   Describe the problem that needs to be solved.
- Scope   Review what this report will and will not cover.

Discussion
- Approach   Describe the proposed solution to the problem.
- Result   Show how the solution will solve the problem.
- Statement of work   List the tasks that will be performed as part of this solution.

Resources
- Personnel   List those who will be doing the work and their qualifications.
- Facilities/equipment   List the physical resources required to do the work.

Costs	
• Fiscal	List the financial costs of implementing the proposed solution.
• Time	List the hours required to implement the proposed solution.
Conclusion	
• Summary	Highlight the benefits and risks of adopting this proposal.
• Contact	Provide a contact for more information.

Introduction

Purpose Tell the reader why you have written this document. The reader may not know or fully understand the purpose. So be specific, and if you are responding to a particular request, say so. For example, what if the purpose were described as follows?

> This document proposes a general system upgrade for ExtremaLab's Sports Analysis Measurement Division (SPASM-D).

Pretty useless! This purpose description does not give the reader adequate specifics to put the proposal in its proper context. Everyone at ExtremaLab is looking for a systems upgrade. In fact, everyone everywhere is looking for a systems upgrade. This proposal might wind up in a pile with everyone else's "wish list." To fix this problem, we need only add specifics:

> This document proposes a general system upgrade for ExtremaLab's Sports Analysis Measurement Division's (SPASM-D) computing capabilities. This upgrade is necessary to enable ExtremaLab's response to the International Olympic Committee's (IOC) tasking for modeling and analysis of the QuadFINKEL figure skating jump.

Background   In this section, describe the problem that needs to be solved, adding any background necessary to clarify the requirement or put it in the proper context. Including specifics also demonstrates your understanding of the problem and adds to your credibility.

> The QuadFINKEL figure skating jump is so demanding that the IOC is considering the award of an extra-large gold medal to anyone successfully landing the jump in Olympic competition. The IOC has tasked SPASM-D, under IOC Contract IOC-135549, to accomplish advanced 3-D modeling, simulation, and analysis of the QuadFINKEL. This modeling and analysis would give the IOC the scientific basis for justifying the special award. This analysis must be completed within 6 months of final project approval.
>
> Modeling this skating jump is a complex process due to the element's chaotic nature and high-speed dynamics. To accomplish this analysis in the specified time frame, SPASM-D, in the attached Technical Report TR-193345, has identified the need for a stand-alone, state-of-the-art graphics, modeling, and analysis capability within the laboratory area. The report identifies the requirement for standard, high-speed TCP/IP interfaces via the Internet to various IOC activities. These stand-alone capabilities must interface through the existing 100/1000 Base-T Fast Ethernet backbone to other laboratory resources. TR-193345 also stipulates that the cost of the upgrade not exceed $22,500.

Scope   Clarify exactly what your proposal covers. *Remember,* an accepted proposal may be considered a binding contract that obligates both parties. Be careful to spell out any exclusions in the scope section. For example, the following scope statement would effectively preclude any responsibility for "add-on" tasks that are not part of the original problem:

This proposal addresses only the system upgrade of the SPASM-D Analysis Laboratory in support of the IOC tasking for modeling and analysis of the QuadFINKEL. This proposal does not include other graphics, modeling, and analysis tasks.

Discussion

Approach   Use this section to describe precisely your proposed solution for the problem. Provide enough details to clearly demonstrate that you have researched the problem, that you understand it, and that you have developed an effective solution.

SPASM-D proposes to the ExtremaLab Technical Review Committee that the SPASM-D laboratory's capabilities be upgraded and augmented with a dedicated local-area network (LAN) composed of two Titanium Graphics 2000 Visual Workstations, one Titanium Graphics 2000A Modeling Station, and one ExtremaLab SuperHUB 1000-X.

Titanium Graphics workstations provide a recognized standard of excellence in performing advanced graphics, modeling, and analysis tasks. Off the shelf, these systems have the required processing power, storage capacity, and network/device interfaces to function seamlessly on the existing network as well as on the Internet. By connecting these workstations with 100/1000 Base-T Fast Ethernet through an ExtremaLab SuperHUB, then cascading this SuperHUB to the company's network backbone, these workstations easily can provide the needed stand-alone computational and storage capabilities. They will also meet all company and Internet connectivity requirements.

Result   Use this section to show what benefits will accrue from the proposed solution. In other words, if ExtremaLab accepts the solution described in the approach section, this is what it will get for its money.

The Titanium Graphics 2000A Modeling Station will provide the dedicated analytical capabilities required to thoroughly model and understand the various components of the QuadFINKEL. The two Titanium Graphics 2000 Visual Workstations will provide the 3-D graphics rendering required by the IOC tasking.

Statement of Work　Use this section to describe the major tasks you will perform to implement the proposed solution. In this example, you might identify three major tasks:

To achieve the goals of this proposal, the following tasks will be accomplished:

- Acquire the necessary equipment; transport, unpack, assemble, and place in the work area. (8 hours)
- Set up the operating systems and configure network connectivity. (16 hours)
- Install application software, check it out, and run calibration and verification simulations. (16 hours)

Resources

Personnel　In this section, discuss who will be doing the work and why they are qualified:

The SPASM-D simulation and modeling staff will analyze and model the QuadFINKEL, using proprietary laboratory software. Additionally, ExtremaLab graphics consultants and IOC figure skating experts will work closely with the simulation and modeling staff to ensure accuracy and effectiveness of all required 3-D rendering.

Facilities and Equipment　Here you should describe the physical resources that will be used to do the work:

The SPASM-D facility includes adequate space for this effort. Suite 104 in Building 45 has the required network access and is available for this project. The specific computer equipment required includes the following:

- Two Titanium Graphics 2000 Visual Workstations*
  - Integrated Visual Computing architecture with Mercury chipset
  - Notel Octagon III 3-gigahertz processors with 2M L-2 cache
  - 4-gigabyte ECC SDRAM
  - 400-gigabyte Ultra2 drive
  - Two 64-bit PCI buses
  - DVD-RW/CD-RW read-write X500 drive
  - Integrated 100/1000 Fast Ethernet
  - IEEE 1394, USB, video, audio ports
  - Winnex IN/IX Open Source OS
  - Three-year warranty with 1-year on-site service
- One Titanium Graphics 2000A Modeling Station
  - Dual Notel Octagon III 3-gigahertz processors with 4M L-2 cache
  - 4-gigabyte ECC SDRAM
  - 500-gigabyte Ultra2 drive
  - Titanium Graphics 6000W 22-inch digital, flat-panel display
  - One ExtremaLab SuperHUB 1000-X
  - Cat 5 100/1000 Base-T cable with connectors

Note: Existing ExtremaLab display monitor resources are available to support these workstations.

Costs

Fiscal Be sure your cost estimate falls within the monetary requirements and constraints of the problem. In this case, the background discussion specified a maximum cost of $22,500.

The proposed system upgrade includes the following equipment and installation costs:

Equipment:

- Two TGI 2000 Visual Workstations
  @ $5995.00 each   = $11,990.00
- One TGI 2000A Modeling Station
  @ $9090.00 each   = $ 9090.00
- One SuperHUB X-1000
  @ $200.00 each   = $ 200.00
- 400 feet Cat 5 100/1000 Base-T
  cable with connectors
  @ $0.19 foot   = $ 76.00

Installation:

- .02 full-time equivalent (FTE)
  technician (40 hours)
  @ $42,000/FTE   = $ 840.00

Total cost: $22,196.00

Time   Be sure your time estimate falls within the requirements and constraints of the problem. In this case, the background discussion specified a maximum time of 6 months to complete the analysis. The schedule for the proposed upgrade has to meet that requirement.

Assuming availability of equipment and materials, and using fully qualified ExtremaLab technicians, the entire upgrade can be completed in 40 hours. This estimate includes 8 hours to acquire, deliver, and set up the equipment; 16 hours to set up the operating system and network connectivity; and 16 hours to configure and check out the applications. This upgrade schedule would provide adequate time to accomplish the modeling and analysis required by the IOC tasking.

Conclusion

Summary   Pay particular attention to this last section. It represents your final opportunity to sell the proposed solution by describing the benefits to be gained through your proposal. This section also allows you to demonstrate your competence by describing how your proposal addresses any risks inherent in the project.

The proposed Titanium Graphics LAN upgrade will provide a viable, cost-effective solution to meeting the IOC QuadFINKEL modeling and analysis requirement. The entire system can be up and running within a week for a price that is well within the cost guidelines. Given the quality of the system and reputation of its manufacturer for setting the standard for high-end analysis and modeling computing systems, the risks of this solution are minimal. In fact, the proposed system provides exactly the right capabilities at precisely the right time, with a cost and time line that are well within the company's needs.

Contact   Tell the reader whom to contact for more information. Be sure that this contact information is accurate and that the person specified understands the proposal and is available to answer questions.

For more information regarding this proposal, contact Edward R. Ronaldson, Ph.D., P.E., Director of Systems Engineering, SPASM-D, at ext. 445; or e-mail: eronald@ExtremaLab.com.

Layout and Presentation

There is an unwritten rule that goes something like this: "Lousy work presented professionally often counts for more than lousy work presented unprofessionally." Many people present foolish ideas so smoothly that they convince others anyway. However, this "style over substance" approach is inappropriate and generally ineffective for technical writing.

You must present your substantive ideas as professionally as possible. Good ideas presented in an unprofessional manner are often interpreted as deficient—especially in technical proposals. To be persuasive, you must present valid ideas clearly and accurately in a document that is professional in every respect, including layout, style, and appearance.

Cover Letters and Title Pages   When submitting a proposal, especially to a recipient external to your organization, you may want to include a *cover letter* (often called a *transmittal letter*). Cover letters ensure that the proposal gets to the right place and is considered in the proper context. These letters provide the following information:

A brief description of the document.

- The problem or requirement that the document. addresses.
- The intended recipient.
- A contact for any additional information.

ExtremaLab
Sports Analysis and Measurement Division (SPASM-D)

March 31, 2004

Dr. Albert M. McEntyre, Chair
ExtremaLab Technical Review Committee
1 Research Plaza, Suite 400
New York, NY 12091

Dear Dr. McEntyre:

Enclosed is the proposal for upgrading SPASM-D modeling and analysis capabilities in response to SPASM-D Technical Report TR-193345 requirements. This upgrade is required to accomplish advanced modeling and analysis of the QuadFINKEL figure skating jump in accordance with International Olympic Committee (IOC) tasking under Contract IOC-135549.

For additional information, please contact Edward R. Ronaldson, Ph.D., Director of Systems Engineering, SPASM-D, at ext. 445, or e-mail: eronald@ExtremaLab.com.

Sincerely,

Evan J. Sonasky
Chief, Modeling and Analysis

EJS/mms
Enclosure: proposal

Additionally, when using a cover letter, you should include a formal title page as the front page of your report. Title pages vary considerably in format, but they generally contain the following information:

- The complete title of the report.
- The name(s) of the author(s).
- The date of submission or period covered by the report.
- The name of the submitting organization.
- The name of the receiving organization.

Title pages may also need to contain additional information, such as RFP numbers and dates, as well as corporate trademarks and designs.

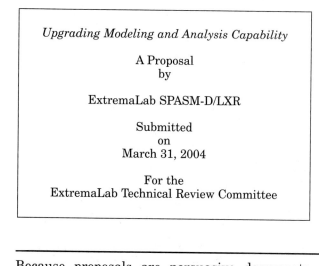

Upgrading Modeling and Analysis Capability

A Proposal
by

ExtremaLab SPASM-D/LXR

Submitted
on
March 31, 2004

For the
ExtremaLab Technical Review Committee

Attachments and Appendixes

Because proposals are persuasive documents designed to sell goods and services, you should not overlook the potential benefit of adding attachments to your report. Such attachments can provide information that supplements or clarifies material in the report but may be too detailed, extensive, or specialized to be included in the report itself. Attachments can be individually numbered and simply attached at the end of the report, or they can be included in a formal appendix section. If placed in an appendix, attachments should be identified individually as Appendix A, Appendix B, and so on.

Attachments can provide a wide range of information. For example, you might use attachments to show specific delivery schedules, detailed cost analysis, or additional background on you or your company. You could even attach additional information that supports your proposal—perhaps a detailed listing of successfully completed projects to establish prior performance. Think about what additional information you can include that will make your proposal more attractive to a potential client.

Putting It All Together

As mentioned, you should consider using cover letters and title pages when sending proposals to recipients external to your organization. When proposals are internal documents, as is the case with our ExtremaLab example, they are often transmitted without a cover letter and title page, using instead an interoffice memo. The document might also be sent through e-mail as a file attachment. When you are using e-mail to transmit a document, always include in the e-mail message the information normally contained in a cover letter, and clearly identify the document in the subject header. Additionally, make sure your e-mail client is using the same attachment encoding as your recipient's client (e.g., MIME, Base64, binhex), and ensure your document file is one that your recipient can use (e.g., Word, pdf, html).

Here is the entire proposal as it might be submitted. Because this document is being sent to an internal recipient, an interoffice memo is used in place of a cover letter and title page.

Date: March 31, 2004
Memo for: The ExtremaLab Technical
Review Committee
From: ExtremaLab SPASM-D/LXR
(E. J. Sonasky)
Subject: Proposal for Upgraded Modeling
and Analysis Capability
Reference: Contract IOC-135549; SPASM-D
TR-193345

Here's the proposal for upgrading SPASM-D
modeling and analysis capabilities. This upgrade
is required to accomplish advanced modeling
and analysis of the QuadFINKEL figure skating
jump in accordance with the referenced contract
and technical report.

For additional information, contact Ed
Ronaldson, SPASM-D, ext. 445, e-mail:
eronald@ExtremaLab.com.

Proposal for Upgrading Modeling
and Analysis Capability

1. Introduction
1.1 Purpose
This document proposes a general system upgrade
for the ExtremaLab's Sports Analysis Measurement Division's (SPASM-D) computing capabilities. This upgrade is necessary to enable
ExtremaLab's response to the International
Olympic Committee's tasking for modeling and
analysis of the QuadFINKEL figure skating jump.

1.2 Background
The QuadFINKEL figure skating jump is so
demanding that the IOC is considering the award

of an extra-large gold medal to anyone successfully landing the jump in Olympic competition. The IOC has tasked SPASM-D, under IOC Contract IOC-135549, to accomplish advanced 3-D modeling, simulation, and analysis of the QuadFINKEL. This modeling and analysis would give the IOC the scientific basis for justifying the special award. This analysis must be completed within 6 months of final project approval.

Modeling this skating jump is a complex process due to the element's chaotic nature and high-speed dynamics. To accomplish this analysis in the specified time frame, SPASM-D, in the attached Technical Report TR-193345, has identified the need for a stand-alone, state-of-the-art graphics, modeling, and analysis capability within the laboratory area. The report identifies the requirement for standard, high-speed TCP/IP interfaces via the Internet to various IOC activities. These stand-alone capabilities must interface through the existing 100/1000 Base-T Fast Ethernet backbone to other laboratory resources. TR-193345 also stipulates that the cost of the upgrade not exceed $22,500.

1.3 Scope

This proposal addresses only the system upgrade of the SPASM-D Analysis Laboratory in support of the IOC tasking for modeling and analysis of the QuadFINKEL. This proposal does not include other graphics, modeling, and analysis tasks.

2. Discussion

2.1 Approach

SPASM-D proposes to the ExtremaLab Technical Review Committee that the SPASM-D laboratory's capabilities be upgraded and augmented with a dedicated local-area network (LAN) composed of two Titanium Graphics 2000 Visual Workstations, one Titanium Graphics 2000A Modeling Station, and one ExtremaLab SuperHUB 1000-X.

Titanium Graphics workstations provide a recognized standard of excellence in performing advanced graphics, modeling, and analysis tasks. Off the shelf, these systems have the required processing power, storage capacity, and network/ device interfaces to function seamlessly on the existing network, as well as on the Internet. By connecting these workstations with 100/1000 Base-T Fast Ethernet through an ExtremaLab SuperHUB, then cascading this SuperHUB to the company's network backbone, these workstations easily can provide the needed stand-alone computational and storage capabilities. They will also meet all company and Internet connectivity requirements.

2.2 Result

The Titanium Graphics 2000A Modeling Station will provide the dedicated analytical capabilities required to thoroughly model and understand the various components of the QuadFINKEL. The two Titanium Graphics 2000 Visual Workstations will provide the 3-D graphics rendering required by the IOC tasking.

2.3 Statement of Work

To achieve the goals of this proposal, the following tasks will be accomplished:

- Task 1: Acquire the necessary equipment; transport, unpack, assemble, and place in the work area. (8 hours)
- Task 2: Set up the operating systems and configure network connectivity. (16 hours)
- Task 3: Install application software, check it out, and run calibration and verification simulations. (16 hours)

3. Resources

3.1 Personnel

The SPASM-D simulation and modeling staff will analyze and model the QuadFINKEL using proprietary laboratory software. Additionally, ExtremaLab graphics consultants and IOC figure skating experts will work closely with the simulation and modeling staff to ensure accuracy and effectiveness of all required 3-D renderings.

3.2 Facilities and Equipment

The SPASM-D facility includes adequate space for this effort. Suite 104 in Building 45 has the required network access and is available for this project. The specific computer equipment required includes the following:

- Two Titanium Graphics 2000 Visual Workstations*
  - Integrated Visual Computing architecture with Mercury chip set
  - Notel Octagon III 3-gigahertz processors with 2M L-2 cache
  - 4-gigabyte ECC SDRAM
  - 400-gigabyte Ultra2 drive
  - Two 64-bit PCI buses
  - DVD-RW/CD-RW read-write X500 drive
  - Integrated 100/1000 Fast Ethernet
  - IEEE 1394, USB, video, audio ports
  - Winnex IN/IX Open Source OS
  - Three-year warranty with 1-year on-site service
- One Titanium Graphics 2000A Modeling Station
  - Dual Notel Octagon III 3-gigahertz processors with 4M L-2 cache
  - 4-gigabyte ECC SDRAM
  - 500-gigabyte Ultra2 drive
  - Titanium Graphics 6000W 22-inch digital, flat-panel display

- One ExtremaLab SuperHUB 1000-X
- Cat 5 100/1000 Base-T cable with connectors

Note: Existing ExtremaLab display monitor resources are available to support these workstations.

4. Costs

4.1 Fiscal

The proposed system upgrade includes the following equipment and installation costs:

Equipment:

- Two TGI 2000 Visual
  Workstations
  @ $5995.00 each   = $11,990.00
- One TGI 2000A Modeling
  Station
  @ $9090.00 each   = $ 9090.00
- One SuperHUB X-1000
  @ $ 200.00 each   = $ 200.00
- 400 feet Cat 5 100/1000
  Base-T cable with connectors
  @ $ 0.19 feet   = $ 76.00

Installation:

- .02 full-time equivalent
  (FTE) technician (40 hours)
  @ $ 42,000/FTE   = $ 840.00

  Total cost: $22,196.00

4.2 Time

Assuming availability of equipment and materials, and using fully qualified ExtremaLab technicians, the entire upgrade can be completed in 40 hours. This estimate includes 8 hours to acquire, deliver, and set up the equipment; 16 hours to set up the operating system and network connectivity; and 16 hours to configure and check out the applications. This upgrade schedule would provide adequate time to accomplish the modeling and analysis required by the IOC tasking.

5. Conclusion

5.1 Summary

The proposed Titanium Graphics LAN upgrade will provide a viable, cost-effective solution to meeting the IOC QuadFINKEL modeling and analysis requirement. The entire system can be up and running within a week for a price that is well within the cost guidelines. Given the quality of the system and the reputation of its manufacturer for setting the standard for high-end analysis and modeling computing systems, the risks of this solution are minimal. In fact, the proposed system provides exactly the right capabilities at precisely the right time, with a cost and time line that are well within the company's needs.

5.2 Contact

For more information regarding this proposal, contact Edward R. Ronaldson, Ph.D., P.E., Director of Systems Engineering, SPASM-D, at ext. 445; or e-mail: eronald@ExtremaLab.com.

The following example is a short, informal proposal that a student might submit to a professor to fulfill a proposal requirement in an engineering technical writing course. In this example, our fictitious student, Sabrina B. McFinkel, is proposing a state-of-the-art research report to satisfy the final course requirement for a formal report on a topic in her major field of study. She is including a cost data section because it is required for the course; however, no one is really going to pay her anything. Additionally, notice the use of first person throughout this document. Such use may be preferred in short, informal documents in which McFinkel's referring to herself in the third person would seem too formal.

A Proposal for a Research Report on the 16XL1000000 Transmitting Tube

By Sabrina B. McFinkel,
EGR 335 student

1. Introduction

1.1 PURPOSE

I propose to research and produce a state-of-the-art report on the 16XL1000000 transmitting tube to fulfill the course requirements of EGR 335, Technical Communication for Engineers and Scientists, Spring 2005.

1.2 BACKGROUND

EGR 335 requires a formal report on a topic in my major field of study (electrical engineering). I am interested in high-power transmitting tubes that are widely used today in RF transmitters supporting TV, FM, and shortwave-broadcast activities as well as military applications in electronic warfare. In fact, high-power RF transmission is one of the few areas of modern electronics in which solid-state devices are not competitive in terms of either cost or efficiency (Anderson 2004, pp. 34–40). I am particularly interested in the 16XL1000000, which is the highest-power transmitting tube currently available in the commercial market, providing continuous RF power outputs of more than a megawatt (XL Tubes).

1.3 SCOPE

My research project will focus primarily on the technical aspects of the 16XL1000000 used in high-power broadcast service. Other aspects of high-power broadcast systems, such as economic, environmental, and political considerations, will not be covered in this report.

2. Discussion

2.1 APPROACH

My proposed state-of-the-art report will begin with a brief review of power tube theory, followed by a short history of RF power amplifier tube design for high-power broadcast service. The report will then focus on the technical aspects of the 16XL1000000. The tube will be analyzed for both class C RF service up to 100 megahertz and AF service in high-level amplitude modulation. Specific areas for investigation will include heat dissipation requirements and techniques, ceramic design considerations, and functional performance characteristics.

2.2 RESULT

My proposed report will provide in-depth technical information on the functional performance of the 16XL1000000 in a variety of broadcast services, and it will include both RF and AF applications in AM, FM, and TV modes. The proposed report will also fulfill the formal report requirement for EGR 335, Technical Communication for Engineers and Scientists.

2.3 STATEMENT OF WORK

To achieve the goals of this research, I will accomplish the following tasks:

1. Briefly review basic theory of RF and AF vacuum tube amplifiers to provide the theoretical context for this report. (3 hours)
2. Research and document RF power amplifier tube design for high-power broadcast service since 1960 to provide a historical context for this device. (5 hours)
3. Research and document the design specifications and manufacturing techniques for the 16XL1000000 that demonstrate the unique characteristics of this device. (5 hours)
4. Describe specific applications for which this device is used and provide examples of the transmission equipment employing this device. (2 hours)

5. Produce and deliver a technical presentation in class at a time and place to be determined. (4 hours)

6. Produce and deliver a formal report document in accordance with the course syllabus requirements. (6 hours)

3. Resources

3.1 PERSONNEL

Consistent with the requirements of the course, I will be doing all the research and writing on this project. I am a senior majoring in electrical engineering with university honors and have a solid background in electromagnetics and electrostatics. I also have a proven track record of success during the past 4 years producing written reports to meet academic requirements. Additionally, I have access to various professors and practicing engineers who can provide any guidance that may be required.

3.2 FACILITIES AND EQUIPMENT

The university provides adequate research and document production facilities for this project. High-speed Internet connectivity, multiple computer systems with all required software, and a number of high-resolution Postscript printers are available for this project.

4. Costs

4.1 FISCAL

As provided in section 2.3 (statement of work), 25 hours of my labor will be required to complete this project. Additionally, 60 miles of travel to and from research facilities will be required, along with the use of the university's computer systems and networks. Access to these networks is supported by the quarterly laboratory fee. The cost of the course itself is represented by tuition charges for 3 credit-hours. The line item cost for the 5-week project ($\frac{1}{2}$ academic quarter) is provided in Table 6.1.

Table 6.1 Project cost data

Labor:	25 hours @ $75.00/hour	=	$1875.00
Travel:	60 miles @ $0.35/mile	=	21.00
Facilities:	Lab fee (pro-rated @ 50% for 5 weeks)	=	75.00
Course:	Tuition (pro-rated @ 50% for 5 weeks)	=	288.00
		Total cost =	**$2259.00**

4.2 TIME

I will accomplish all tasks during the last 5 weeks of the quarter and will provide all deliverables according to the following schedule:

- Presentation: At a time and place to be determined by the course instructor.
- Report: In the classroom, at 9:30 a.m. on Thursday, June 2, 2005.

5. Conclusion

5.1 SUMMARY

EGR 335 requires a formal report document and formal report presentation on a topic in electrical engineering, my major field of study. The 16XL1000000 transmitting tube is the highest-power RF amplifier available today and is used in a variety of broadcast and military applications. It represents the state of the art in high-power transmission devices and, as such, provides an appropriate topic for research in my major. Approval of this proposal will allow me to fulfill the course requirements while learning more about high-power broadcast systems.

5.2 CONTACT

For more information, please e-mail me at mcfinkel@wxccs.edu, or phone (937) 999-1234.

5.3 SOURCES USED

Anderson, Warren S.: *Big Tubes—Big Power.* New York: RF Press, 2004.
XL Tubes, Internet: http://www.support.xltubes.com/16XL1000000.html, May 2003.

Proposal Checklist	• Have I defined the problem in great enough detail to ensure that my reader will understand the context for this proposal?
	• Have I described the background to this problem in great enough detail to clearly identify the variables driving my proposed solution?
	• Have I defined in the scope section how I am limiting my proposal?

- Have I laid out my proposed solution in adequate detail? Do I have enough details to ensure that my solution is credible?
- Have I clearly described the benefits of my solution?
- Do I need to provide a statement of work? If so, have I adequately described the major tasks involved in implementing the proposed solution?
- Have I clearly defined the resources required to implement my proposed solution, including people, facilities, and equipment?
- Have I provided the required budget for implementing this proposal, including the costs in both money and time? Have I broken out the costs in adequate detail for my audience? Are these cost estimates consistent with the financial constraints of the problem?
- Is my time estimate consistent with the tasks in my statement of work?
- Have I clearly summarized the proposal and provided a strong concluding argument for its adoption?
- Have I provided an available and knowledgeable contact at the end of the proposal?

CHAPTER SEVEN

Progress Reports

Once a proposal has been accepted and work begins on a project, you may be required to give the client (or your professor, as the case may be) periodic reports on how well you are doing. A *progress report* (often called a *status report*) is the type of document commonly used to detail and document the status of the project. Writing a progress report usually means answering straightforward questions such as, Are you on schedule? Are you within budget? Are any risks evident? If so, how do you plan to control them? What problems have you encountered? What remains to be done? What is your plan for doing it? What is your overall assessment?

Does the progress report requirement indicate that your client or professor does not trust you? Maybe—but that is not the point. In the "real world," normal business practice requires specific, written documentation, not abstract trust. In some cases, continuation of the contract or partial payments for work completed (or even a grade in a course) may depend on your writing and submitting satisfactory progress reports.

On larger projects, progress or status reports may also be known as *milestone reports,* because, like milestones on a road, they mark the passage from one point to another on a journey toward some final goal. In the real world, periodic payments may be contingent on your submitting successful milestone reports. Additionally, you may be asked by your boss periodically to submit an *activity report.* This type of project report provides the ongoing status of a project (or projects) for which you are responsible—kind of a "tell me what you have done for me lately" report.

Progress reports, status reports, milestone reports, and activity reports all do basically the same thing and contain the same kinds of information. So to keep things simple, this chapter will focus only on progress reports.

What Is a Progress Report?

Progress reports document the status of a project. These reports describe the various tasks that make up the project and analyze the progress that has been made toward completing each task. If you have written a proposal that has been accepted, then you also have probably committed yourself (or your organization) to completing certain tasks by a particular time and for a specified amount of money. Generally speaking, in a progress report you need to tell the reader three things: the problem you are solving, the solution you are implementing, and how well you are doing. You will find that writing a progress report is more pleasant when you have some progress to report.

Writing a progress report typically requires that you do three things: review, describe, and evaluate.

1. *Review* the problem that was the impetus for the original proposal. To do this, reference the original proposal by number or title, indicate when it was accepted, and then describe the problem that prompted the proposal in the first place. Often you can copy much of the problem description from the original proposal into the progress report.

2. *Describe* the solution offered in the original proposal, including the tasks involved in implementing this solution (usually listed in the statement of

work) and the planned dates for completing each task.

3. *Evaluate* how well you are doing in terms of each task, and provide an overall assessment of your progress. In other words, lay out to what extent you are getting the job done within the time frame and cost constraints of the original proposal.

Progress Report Formats

Progress reports can take many forms, from a simple letter to a multipage document. Use the format that your boss or instructor tells you to use, or the one you are required to use by the client. If you do not have a specific format, Outline 7.1 should do the trick.

To better understand how progress reports work, consider each part of this outline and how you would use it. This example follows up Chapter 6's proposal for setting up the local-area network to analyze and model the QuadFINKEL figure skating jump. Assume that the proposal was accepted and the project is under way.

The ExtremaLab Technical Review Committee has asked for a short progress report that can be faxed to the International Olympic Committee for a meeting the day after tomorrow. No big deal! Producing this type of report is not difficult, especially if you have a copy of the original proposal handy. You will need that original proposal because it describes the context and tasks on which you will now report your organization's progress.

Introduction
Purpose
Following Outline 7.1, first open the introduction by stating the purpose of this progress report. Think about the purpose carefully. Writing an effective report can be difficult if you cannot articulate why you are doing it.

This report documents the progress ExtremaLab has made on upgrading the modeling and analysis capability pursuant to its accepted SPASM-D/LXR proposal of March 31, 2004.

Outline 7.1   Progress Reports

Introduction
* Purpose   Review the reason for writing this report.
* Background Review the problem and your
              proposed solution.
* Scope   Describe what this report will and will not cover.

Status

Tasks completed
* For each relevant task (repeat this pattern for each task or activity completed)
  * Provide a description of the task.
  * Describe the things that have been accomplished.
  * Describe how long it took to accomplish them.
  * Describe the difficulties, if any, that were encountered.

Tasks remaining
* For each relevant task (repeat for each task or activity remaining)
  * Provide a description of the task.
  * Describe the things that still need to be accomplished.
  * Provide the timetable and strategy for completing the task.
  * Describe the risks and your approach for completing the task.

Conclusion
* Summary   Provide an appraisal of the current status.
* Evaluation Provide an assessment of the progress made to date.
* Forecast   Provide a forecast for completing this project.
* Contact   Provide a contact for more information.

Background

In the background section of the introduction, describe the context for the project on which you are reporting. The best approach is to reference the accepted proposal, note its date of acceptance, and describe the specific problem it addressed, along with the proposed solution.

The QuadFINKEL figure skating jump is so demanding that the IOC has tasked SPASM-D, under IOC Contract IOC-135549, to accomplish advanced 3-D modeling, simulation, and analysis of the QuadFINKEL. Modeling this skating jump is a complex process due to the jump's chaotic nature and high-speed dynamics.

To accomplish this analysis in the specified time frame, SPASM-D proposed to the ExtremaLab Technical Review Committee that the SPASM-D laboratory's capabilities be upgraded and augmented with a dedicated local-area network (LAN). The new LAN will be composed of two Titanium Graphics 2000 Visual Workstations, one Titanium Graphics 2000A Modeling Station, and one ExtremaLab SuperHUB 1000-X. The proposal was accepted without change on June 1, 2004, with the total system upgrade to be completed on or before June 30, 2005.

Scope

In the scope section of the introduction, describe the tasks or aspects of the project that this progress report covers. If the original proposal included a statement of work, you will probably report on some of or all the tasks listed. Specify what tasks you will be reporting on in this progress report. If applicable, specify which tasks you have reported on in an earlier report, which tasks you will report on in a subsequent report, and whether any tasks are recurring. If the original proposal did not include a statement of work, you will need to provide a more general description of the tasks on which you are reporting.

This report provides the status of all tasks described in the accepted proposal's statement of work, which includes the following:

- Acquiring the necessary equipment; transporting, unpacking, assembling, and placing in the work area. (8 hours)

- Setting up the operating systems and configuring network connectivity. (16 hours)
- Installing application software, checking it out, and running calibration and verification simulations. (16 hours)

Status

This section presents the status of each task listed in the scope section. Normally, you would treat completed tasks separately from the remaining tasks. The tasks-completed section includes tasks that have been concluded and closed out. The tasks-remaining section includes tasks that are still in progress and tasks that have not yet been started. Write the tasks-completed section first, discussing each task separately:

Tasks Completed

Task 1: Acquiring the necessary equipment; transporting, unpacking, assembling, and placing in the work area. (8 hours)
- All necessary equipment—including two Titanium Graphics 2000 Visual Workstations, one Titanium Graphics 2000A Modeling Station, and one SuperHUB 1000-X—was purchased on June 2, 2004, within budget estimates, and delivered, unpacked, and placed in Suite 104, Building 45, on June 4, 2004. The entire task required about 7 hours. No difficulties were encountered.

Task 2: Setting up the operating systems and configuring network connectivity. (16 hours)
- All three workstations came with the Winnex IN/IX Open Source OS preinstalled. We successfully configured these machines and the hub to operate on the LAN. Initially the LAN experienced an excessive number of packet collisions. The problem was isolated to the SuperHUB, which was subsequently replaced. We completed system setup and network configuration in 8 hours.

Once you've finished the tasks-completed section, write the tasks-remaining section in the same way:

Tasks Remaining

Task 3: Installing application software, checking it out, and running calibration and verification simulations. (16 hours)

- We have just started this task. We are well ahead of schedule on implementing the statement of work, and we foresee no problems in completing this task and the entire project on time and within specified budget constraints.

Conclusion

The conclusion of the progress report summarizes and appraises the progress to date and provides a forecast for the rest of the project (including any risks and plans for their mitigation). It also provides a contact in case more information is required. As in the proposal, the contact should be someone who is familiar with the project.

The SPASM-D technical staff is pleased with the progress made to date. We procured, transported, unpacked, assembled, and installed the equipment in its intended operating location in only 7 hours, thereby completing task 1 ahead of schedule. We configured the operating systems and achieved full network connectivity in 15 hours, thereby completing task 2 on schedule. We are now in the process of installing, checking out, and calibrating the application software and expect to complete task 3 within the scheduled 16 hours. Consequently, we at ExtremaLab/ SPASM-D assess the progress on this project as *excellent* and are confident that the entire project will be completed on time and within the proposed budget.

For more information on this project, contact Edward R. Ronaldson, Ph.D., P.E., Director of Systems Engineering, SPASM-D, at ext. 445; or e-mail: eronald@ExtremaLab.com.

Putting It All Together

Here is the entire progress report assembled as it might be submitted. Notice the addition of Figure 7.1, a time line chart showing the relative progress for the three tasks. For short, informal reports, a time line chart normally is not required; one is included here only as an example. Also notice that a memorandum header is used without a title page or cover letter since this progress report is an informal, internal document. If it were designed for use outside the company, it would need a cover letter and title page (Chapter 6).

Date: June 7, 2004

To: The ExtremaLab Technical Review Committee

From: ExtremaLab SPASM-D/LXR

Subject: Progress Report for Upgraded Modeling and Analysis Capability

Reference: ExtremaLab Proposal #CL-3478 of March 31, 2004

1. Introduction

1.1 Purpose

This report documents the progress ExtremaLab has made on upgrading the modeling and analysis capability pursuant to its accepted SPASM-D/LXR proposal of March 31, 2004.

1.2 Background

The QuadFINKEL figure skating jump is so demanding that the IOC has tasked SPASM-D, under IOC Contract IOC-135549, to accomplish advanced 3-D modeling, simulation, and analysis of the QuadFINKEL. Modeling this skating jump is a complex process due to the jump's chaotic nature and high-speed dynamics. To accomplish this analysis in the specified time frame, SPASM-D proposed to the ExtremaLab Technical Review Committee that the SPASM-D laboratory's capabilities be upgraded and augmented with a dedicated local-area network (LAN). The new LAN will be composed of two Titanium Graphics 2000 Visual Workstations, one Titanium Graphics 2000A Modeling Station, and one ExtremaLab SuperHUB 1000-X. The proposal was accepted without change on June 1, 2004, with the total system upgrade to be completed on or before June 30, 2005.

1.3 Scope

This report provides the status of all tasks described in the accepted proposal's statement of work, which includes the following:

- *Task 1:* Acquiring the necessary equipment; transporting, unpacking, assembling, and placing in the work area. (8 hours)
- *Task 2:* Setting up the operating systems and configuring network connectivity. (16 hours)
- *Task 3:* Installing application software, checking it out, and running calibration and verification simulations. (16 hours)

2. Status

2.1 Tasks Completed

Task 1: Acquiring the necessary equipment; transporting, unpacking, assembling, and placing in the work area. (8 hours)

- All necessary equipment—including two Titanium Graphics 2000 Visual Workstations, one Titanium Graphics 2000A Modeling Station, and one SuperHUB 1000-X—was purchased on June 2, 2004, within budget estimates, and delivered, unpacked, and placed in Suite 104, Building 45, on June 4, 2004. The entire task required about 7 hours. No difficulties were encountered.

Task 2: Setting up the operating systems and configuring network connectivity. (16 hours)

• All three workstations came with Winnex IN/IX Open Source OS preinstalled. We successfully configured these machines and the hub to operate on the LAN. Initially the LAN experienced an excessive number of packet collisions. The problem was isolated to the SuperHUB, which was subsequently replaced. We completed system setup and network configuration in 8 hours.

2.2 Tasks Remaining

Task 3: Installing application software, checking it out, and running calibration and verification simulations. (16 hours)

• We have just started this task. We are well ahead of schedule on implementing the statement of work, and we foresee no problems in completing this task and the entire project on time and within specified budget constraints.

3. Conclusion

The SPASM-D technical staff is pleased with the progress made to date (see Figure 7.1 for the project time line). We procured, transported, unpacked, assembled, and installed the equipment in its intended operating location in only 7 hours, thereby completing task 1 ahead of schedule. We configured the operating systems and achieved full network connectivity in 15 hours, thereby completing task 2 on schedule. We are now in the process of installing, checking out, and calibrating the application software and expect to complete task 3 within the scheduled 16 hours. Consequently, we at ExtremaLab/SPASM-D have assessed the progress on this project as *excellent*

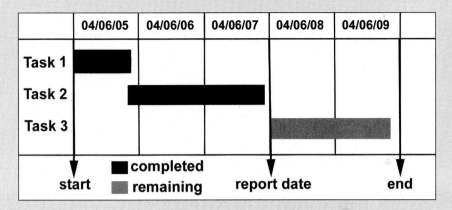

Figure 7.1
Project time line.

and are confident that the entire project will be completed on time and within the proposed budget.

For more information on this project, contact Edward R. Ronaldson, Ph.D., P.E., Director of Systems Engineering, SPASM-D, at ext. 445; or e-mail: eronald@ExtremaLab.com.

Student Progress Report

The following example is a short, informal student progress report required by an instructor for an assigned project. In this case, Sabrina B. McFinkel is reporting on the status of her state-of-the-art report on the 16XL1000000 transmitting tube (see her original proposal in the Threaded Example of Chapter 6).

> **Memo for:** Albert J. Finnian, Ph.D., Instructor, EGR 335
> **From:** Sabrina B. McFinkel, EGR 335 student
> **Subject:** Progress of class project
> **Reference:** Proposal for research report on 16XL1000000, April 5, 2004

1. Introduction

1.1 PURPOSE

This report provides the current status of my research report.

1.2 BACKGROUND

On April 5, 2005, I proposed a state-of-the-art research report on the 16XL1000000 transmitting tube. You accepted my proposal as submitted on April 8, 2005. We agreed that my proposed report would provide in-depth technical information on the functional performance of the 16XL1000000 in a variety of broadcast services, and include both RF and AF applications in AM, FM, and TV modes. The proposed report would also fulfill the formal report requirement for EGR 335, Technical Communication for Engineers and Scientists.

1.3 SCOPE

This status report provides my current assessment of the project as of Monday, May 2, 2005. The statement of work included the following tasks:

- Task 1: Briefly review basic theory of RF and AF vacuum tube amplifiers to provide the theoretical context for this report. (3 hours)
- Task 2: Research and document RF power amplifier tube design for high-power broadcast service since 1960 to provide historical context for this device. (5 hours)
- Task 3: Research and document the design specifications and manufacturing techniques for the 16XL1000000 that demonstrate the unique characteristics of this device. (5 hours)
- Task 4: Describe specific applications for which this device is used and provide examples of the transmission equipment employing this device. (2 hours)
- Task 5: Produce and deliver a technical presentation in class at a time and place to be determined. (4 hours)

- Task 6: Produce and deliver a formal report document in accordance with the course syllabus requirements. (6 hours)

2. Status

2.1 TASKS COMPLETED

2.1.1 *Task 1:* I successfully completed this review on April 11, 2005, without any problems.

2.1.2 *Task 2:* I successfully completed this task on April 16, 2005. Initially, I had difficulty finding adequate sources; however, I subsequently located the necessary information on a foreign manufacturer's website.

2.2 TASKS REMAINING

2.2.1 *Task 3:* This task is partially complete. I received design specifications from the manufacturer on April 21, 2005; however, I was advised that the manufacturing techniques are proprietary and would not be available. I now plan to research nonproprietary manufacturing techniques of similar tubes to address this task, and I expect to complete the task by May 9, 2005.

2.2.2 *Task 4:* I have not completed this task, although I have located adequate research materials and will complete this task by May 9, 2005.

2.2.3 *Task 5:* The presentation has been scheduled for May 19, 2005, but I will not produce the briefing charts until tasks 3 and 4 are complete. The current schedule allows me to delay this task and still present the briefing as scheduled.

2.2.4 *Task 6:* I have already drafted about half of this report and will complete the initial draft once I have finished tasks 3 and 4. I will work on this task in parallel with task 5 and will deliver the finished state-of-the-art report as scheduled on Thursday, June 2, 2005.

3. Conclusion

3.1 ASSESSMENT

I am pleased with the project to date and assess my progress as *excellent*. Most of the major research has been completed, and I am making good progress toward filling in the gaps where information is still lacking. The balance of task 3 and all of task 4 should be completed by May 9, 2005. After that, I will complete task 5 and deliver the presentation as scheduled on May 19, 2005. I also have more than adequate time to produce the final deliverable, state-of-the-art report, and I will deliver it as required by Thursday, June 2, 2005.

3.2 CONTACT

For more information, please e-mail me at mcfinkel@wxccs.edu, or phone (937) 999-1234.

Progress Report Checklist

- Have I specified the purpose, background, and scope of this report?
- Have I referenced the accepted proposal by name, number, and/or date?
- Have I reviewed the problem contained in that proposal?
- Have I reviewed the proposed solution to that problem?
- Have I specified the tasks that will be included in this report?
- Have I properly discussed the tasks completed and tasks remaining?
- Have I provided an appraisal and forecast in the conclusion?

08

Feasibility and Recommendation Reports

Feasibility reports and *recommendation reports* are objective documents that identify and evaluate solutions to problems. In technical writing, these reports address subjects that have well-defined parameters, including a problem, or multiple problems, that can be precisely described; and solutions, or multiple solutions, that can be objectively and empirically tested.

Feasibility and recommendation reports are unbiased evaluations, although their conclusions and recommendations can be—and frequently are—used to promote ideas and sell goods and services. Ideally, however, only someone who is totally impartial should write these reports. The author should have no stake in the outcome and should not care whether any of or all the solutions are adopted.

How Do Feasibility and Recommendation Reports Differ?

Feasibility reports and *recommendation reports* are similar, and the terms are often used synonymously. Both reports define a problem and objectively evaluate solutions based on a set of criteria—but with one small difference. Feasibility reports, on one hand, determine the feasibility or viability of solving a problem in a particular way. In other words, feasibility reports consider a single solution to a problem and determine whether

125

or not, or to what extent, the solution is feasible. Recommendation reports, on the other hand, look at several approaches for solving a problem and recommend the most feasible approach. Both feasibility and recommendation reports, then, basically do the same thing: They objectively evaluate the feasibility of solutions.

Outline 8.1 provides a typical approach for organizing a feasibility report. Outline 8.2 does the same thing for a recommendation report.

Outline 8.1 Feasibility Report

Introduction
- Purpose   Describe the reason for writing this report.
- Problem   Describe the problem that needs to be solved.
- Scope   Describe the proposed solution and list criteria.

Discussion
- Criterion 1
  - Explanation   Describe the criterion, why it was selected, and how it is used.
  - Data   Provide the findings (data) for this criterion.
  - Interpretation Interpret the data for this criterion relative to the solution.
- Criteria 2–*n*
(Treat each remaining criterion in the same manner as criterion 1.)

Conclusion
- Summary   Review the data and interpretations.
- Conclusions   Present your conclusions based on the data and interpretations.
- Recommendation   Recommend or reject the proposed solution.

Contact   Provide a contact for this report.

Documentation   List the sources and references used.

Appendix
- Include materials needed for support, but not for understanding the report.

> ## Outline 8.2 Recommendation Report
>
> Introduction
> * Purpose   Describe the reason for writing this report.
> * Problem   Describe the problem that needs to be solved.
> * Scope   Describe the proposed solutions and criteria.
>
> Discussion
> * Criterion 1
>   * Explanation   Describe the criterion, why it was selected, and how it is used.
>   * Data   Provide the findings (data) for this criterion for each solution.
>   * Interpretation Interpret the data for this criterion relative to each solution.
> * Criteria 2–n
>   (Treat each remaining criterion in the same manner as criterion 1.)
>
> Conclusion
> * Summary   Review the data and interpretations.
> * Conclusions   Present your conclusions based on the data and interpretations.
> * Recommendation   Recommend the best solution based on these conclusions.
>
> Contact   Provide a contact for this report.
> Documentation   List the sources and references used.
> Appendix
> * Include materials needed for support, but not for understanding the report.

As Outlines 8.1 and 8.2 show, you need to do several things when writing either a recommendation or a feasibility report:

Writing Feasibility and Recommendation Reports

1. *Define a problem that needs to be solved.* The difficulty here is that we are often solution-oriented; in many cases we skip the problem and go directly to the solution. What if a friend came to you and said, "I have a problem: I need to buy a computer, but I do not know which one to purchase"? The main problem your friend has is that he or she does not have a problem. "I need

to buy a computer" states a solution, not a problem. Why does your friend need a computer? That is the problem! To come up with candidate solutions, you would have to know why your friend needs a computer. Feasibility and recommendation reports work the same way. You cannot evaluate a solution to a problem that is not clearly defined.

2. *Identify one or more candidate solutions.* This process can be tough; sometimes many more solutions exist than you will have the time or capability to evaluate. If you need a computer to surf the Internet, how many choices do you have? This is almost like asking how many stars are in the Milky Way galaxy. Coming up with just a few viable solutions can be challenging. Normally, you can apply additional requirements to the existing problem that will allow you to narrow the list. Maybe you will buy only from an approved, local vendor; or you will shop only within 5 square miles of your home; or you will consider using only a certain catalog that gives you a rebate.

3. *Develop a set of criteria by which to objectively evaluate the candidate solution(s).* The key here is *objective.* Find meaningful measures that relate to the problem you have defined, and identify valid methods for applying them. For example, when looking for a computer to surf the Internet, you might use criteria such as cost, processor speed, monitor size and quality, reliability and warranty, bundled software, and included peripherals. These can be objectively described and measured. The attractiveness of the case would not be a good criterion because computer case attractiveness cannot easily be objectively described and measured.

4. *Collect data for each criterion as it relates to each candidate solution, and then interpret those data.* One thing that must be decided is how to weight the importance of each criterion. Sometimes the criteria can be weighted equally,

but in many studies, some criteria are more important and need to count more in the final decision. For example, what if you wanted a computer to surf the Internet using a satellite link from the back of an all-terrain vehicle in the upper Amazon region? In this case, reliability and maintainability might be far more important than, say, processor speed. However, if you planned to use the computer to do serious number crunching in the office, processor speed would be more important.

5. *Draw conclusions and make recommendations regarding the feasibility of the candidate solutions based on your interpretations.* The primary requirements here are objectivity and clear thinking. Look at the interpretations you have made for each criterion, and consider the relative weighting of the criteria. Always base your conclusions on this information—never on other information or considerations that are not fully treated in the report.

Both feasibility and recommendation reports are organized and written in the same straightforward, logical manner. To simplify this discussion, a recommendation report example will be developed first, since it includes all the elements of a feasibility report. This recommendation report example identifies and evaluates, in the context of astronautics, two alternative transfer maneuvers for moving a fictitious signals reconnaissance satellite (code-named "Big Ears") from one earth orbit to another. This type of recommendation report might be used as an internal position or background paper, which is also referred to frequently as a *white paper.*

Introduction
Purpose
Describe the purpose of this report. This part is straightforward. What do you hope to achieve with this document? To what requirement does

this document respond? Your purpose statement might look something like this:

> The purpose of this report is to recommend to NASA the best orbital transfer maneuver for moving a satellite from a low equatorial orbit to a high geosynchronous orbit.

Once you have stated the purpose, you can define the problem addressed in this report.

Problem

You cannot write a feasibility report unless you have a problem. Solutions without problems are not very useful, and neither are solutions that are not geared specifically to solving problems. So to define the problem in this section, you should provide the needed background and describe specific requirements that any viable solution must meet.

Consider the following problem statement:

> The National Intelligence Authority needs to move a satellite from its present near-earth orbit to a geosynchronous orbit.

This problem statement is not very useful because it does not include adequate information about the problem. How can anyone possibly identify and evaluate a solution if the problem is not properly defined in the first place? In fact, in this case, moving the satellite is actually part of the solution. The "what," "when," "why," and "how" are important aspects of the problem that need to be addressed. To describe the problem, then, you need to specify the kind of satellite that needs to be moved, why it needs to be moved, how much time is available to get it where it needs to go, and how accurate the transfer maneuver needs to be. A better problem statement might look something like the following:

NASA recently launched a Big Ears reconnaissance satellite from Cape Canaveral and injected it into a 300-kilometer near-earth orbit (NEO) over the equator (0° inclination). The satellite has been successfully configured and checked out and now needs to be moved to its operational position in geosynchronous earth orbit (GEO) at an altitude of 38,786 kilometers with an inclination of 0°. This position will enable the satellite to remain stable over the earth's surface, so that it can survey those portions of the earth required by its mission profile. The National Intelligence Authority requires that the satellite be on station by 0500 GMT on June 1, 2005. To allow time for full calibration on orbit, an appropriate transfer maneuver must be selected and executed at least 72 hours prior to this time. Since the satellite is already in an equatorial orbit, transferring the satellite from the current orbit to the required orbit involves only an altitude change within the same plane. Two possible maneuvers are being considered to accomplish this transfer: the Hohmann transfer (HT) and the one-tangent burn (OTB). The HT is a doubly tangent, transfer ellipse that requires two tangent burns—one at NEO and one at GEO. The OTB requires a single-tangent burn at NEO and a second nontangent burn at GEO.

Scope

The scope section lists the candidate solutions to be considered, along with the criteria that will be used to test each solution. First ensure that any candidate solution you consider is at least apparently viable. Some solutions may look good initially; however, when they are closely examined in terms of the problem to be solved, these same ideas may appear silly. Take a preliminary look at each candidate solution to ensure it makes sense.

Also list the criteria by which you will evaluate each candidate solution. These criteria, driven by the problem, can include almost anything from cost, to reliability, to ease of use. Normally, in a technical paper, you use only criteria that can be objectively and empirically quantified and tested. In the case of our example problem, we know that the solution must be robust, reliable, economically feasible, and efficient. Based on these requirements, we might develop a scope section that looks like the following:

> This report compares the HT to the OTB, using three criteria: accuracy, time of flight, and required fuel.

Discussion

The discussion section is the heart of the feasibility report. In this section you discuss the criteria, apply them to the candidate solutions, generate data, and interpret the results. Notice that this section is organized by criteria, not solutions. That is extremely important. You want to compare solutions for each criterion, not vice versa. If instead you organize the discussion around solutions, you will have to go through all your criteria for each solution. By the time you get through all the solutions, you will find it difficult to conceptualize and compare the solutions effectively because you will not remember which data pertained to which solution.

To begin the discussion section, here is the kind of treatment you might give the first criterion, *accuracy*.

Accuracy

First, in the explanation section, define the criterion, explain why you selected it (if necessary), and note any special weighting it will have in the evaluation process. If it has special weighting, explain how you chose your weighting system.

> *Explanation:* Accuracy refers to the relative precision with which the satellite is placed into its target orbit. If a satellite is not accurately placed, additional

onboard fuel will be required to adjust the orbit. Normally, a transfer is considered to be accurate if the radius of the target orbit is within ±0.001 DU (1 DU = .5 earth diameter, or 6378.145 kilometers). Because the problem is relatively short term, no special weighting is assigned to this criterion. Onboard thrusters can correct for insertion errors; however, the consequent use of fuel may have long-term implications.

In the data section, discuss the findings, using this criterion for each alternative solution. Be sure to properly document the source for the data being used.

Data: Using this criterion, and in terms of past performance, the HT has achieved an accurate transfer 85.5 percent of the time. The accuracy for the OTM is at 60.2 percent. (Orbitus 2004, pp. 224–228)

In the interpretation section, discuss your interpretation of the data for this criterion for each alternative solution.

Interpretation: In terms of past performance, the HT provides higher accuracy than the OTB. Through the use of onboard thrusters, either maneuver can solve the current problem; however, the greater accuracy of the HT has positive, long-term implications for system viability on-station. The energy to precisely position a satellite on station is provided by a finite supply of onboard fuel, which also will be used to power the onboard thrusters for later, station-keeping activities on orbit. More accurate insertion means less fuel required for positioning, thus conserving more fuel for subsequent station-keeping functions. Since higher accuracy translates to longer operational life for the satellite, the HT is preferred.

Next, in a similar manner, deal with the second criterion, *time of flight.*

Time of Flight

Describe the criterion, explain why you selected it, and note any special weighting it will have in the evaluation process.

> *Explanation:* Time of flight (TOF) is the elapsed time between the initial firing of the satellite's retro-rockets at a tangent to the initial NEO and the point when the satellite intersects the radius of its target GEO. Because the problem does not provide time-critical constraints, this criterion is given no special weighting.

Next, provide the data.

> *Data:* Using the HT, the satellite requires a TOF of 24.10 TU (1 TU = 13.44686457 minutes). The OTB provides a TOF of 23.45 TU to transfer the satellite from NEO to GEO. (See Appendix A for the calculations supporting these results.)

Discuss your interpretation of the data for this criterion for each alternative solution.

> *Interpretation:* The OTB requires less time than the HT to transfer the satellite from NEO to GEO. Because time-sensitive constraints do not exist in this case, and because the differences in TOF for both options are relatively small, no preference exists for either choice.

Required Fuel

Handle this criterion in a manner similar to that for the first two.

> *Explanation:* As mentioned earlier, the energy to move the satellite from NEO to GEO is provided by onboard fuel, which also powers the onboard thrusters for station-keeping activities on orbit. Fuel not required for the orbital transfer would then be available for future on-orbit maneuvers.

The satellite moving in NEO possesses considerable kinetic energy, as shown in Equation (1).

$$E = \frac{1}{2}mv^2 \qquad (1)$$

While the mass m of the satellite is considered to be constant, the change in velocity v requires an increase in energy which, in turn, requires more fuel.

Data: The total change in velocity for the HT is .498 DU/TU (1 DU/TU = 7.90536828 km/s). The total change in velocity for the OTB is .594 DU/TU. (See calculations in Appendix B.)

Interpretation: The HT requires a smaller change in velocity, thereby requiring less fuel compared to the OTB. In terms of fuel savings, the HT is preferred.

Conclusion

In the conclusion section, summarize the data and interpretations for all criteria and solutions. Do not include new information in the conclusion section; it should deal only with information already presented in the paper.

Summary

Briefly summarize the report and its findings.

This study evaluated both the HT and OTB based on three criteria: accuracy, time of flight, and fuel required. The HT's performance was better in terms of accuracy and required fuel. The OTB's performance was better in terms of TOF.

Conclusion

Provide any conclusion you can draw from the summary above.

The HT is preferred based on accuracy and fuel requirements. The OTB performed better in terms of time of flight, but this advantage is not considered significant given the constraints of the problem.

Recommendation

Make your recommendation.

The HT is recommended as the transfer maneuver for moving the Big Ears satellite from NEO to GEO.

Contact

Finally, provide any needed contact information.

For more information, contact James R. McFinkel, Ph.D., NIA/OR3, at (202) 444-9999.

Putting It All Together Here is the recommendation report, assembled, formatted, and modified to include supporting diagrams and a summary table of data. The extensive calculations that would comprise the appendixes have not been included in this chapter because they are not needed for the purposes of this example.

Recommendation Report on Transfer Maneuvers for the Big Ears Satellite

Introduction

Purpose

The purpose of this report is to recommend the best orbital transfer maneuver for moving a satellite from a low equatorial orbit to a high geosynchronous orbit.

Problem

NASA recently launched a Big Ears reconnaissance satellite from Cape Canaveral and injected it into a 300-kilometer near-earth orbit (NEO) over the equator (0° inclination). The satellite has been successfully configured and checked out, and now it needs to be moved to its operational position in geosynchronous earth orbit (GEO) at an altitude of 38,786 kilometers with an inclination of 0°. This position will enable the satellite to remain stable over the earth's surface so that it can survey those portions of the earth required by its mission profile.

The National Intelligence Authority requires that the satellite be on station by 0500 GMT on June 1, 2005. To allow time for full calibration on orbit, an appropriate transfer maneuver must be selected and executed at least 72 hours prior to this time. Since the satellite is already in an equatorial orbit, transferring the satellite from the current orbit to the required orbit involves only an altitude change within the same plane. Two possible maneuvers are being considered to accomplish this transfer: the Hohmann transfer (HT) and the one-tangent burn (OTB). The HT is a double-tangent transfer ellipse that requires two tangent burns—one at NEO and one at GEO (Figure 8.1a). The OTB requires a single-tangent burn at NEO and a second nontangent burn at GEO (Figure 8.1b).

Scope

This report compares the HT to the OTB, using three criteria: accuracy, time of flight, and required fuel.

Discussion

Accuracy

Explanation Accuracy refers to the relative precision with which the satellite is placed into its target orbit. If a satellite is not accurately placed, additional onboard fuel will be required to adjust the orbit. Normally, a transfer is considered to be accurate if the radius of the target orbit is within ±0.001 DU (1 DU = .5 earth diameter, or 6378.145 kilometers). Because the problem is relatively short term, no special weighting is assigned to this criterion. Onboard thrusters can

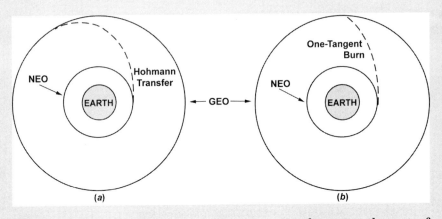

Figure 8.1
(*a*) Hohmann transfer;
(*b*) one-tangent burn.

correct for insertion errors; however, the use of fuel may have long-term implications.

Data   Using this criterion, and in terms of past performance, the HT has achieved an accurate transfer 85.5 percent of the time. The accuracy for the OTM is at 60.2 percent. (Orbitus 2004, pp. 224–228)

Interpretation   In terms of past performance, the HT provides higher accuracy than the OTB. Through the use of onboard thrusters, either maneuver can solve the current problem; however, the greater accuracy of the HT has positive long-term implications for system viability on-station. The energy to precisely position a satellite on station is provided by a finite supply of onboard fuel, which also will be used to power the onboard thrusters for later station-keeping activities on orbit. More accurate insertion means less fuel required for positioning, thus conserving more fuel for subsequent station-keeping functions. Since higher accuracy translates to longer operational life for the satellite, the HT is preferred.

Time of Flight

Explanation   Time of flight (TOF) is the elapsed time between the initial firing of the satellite's retrorockets at a tangent to the initial NEO and the point when the satellite intersects the radius of its target GEO. Because the problem does not provide time-critical constraints, this criterion is given no special weighting.

Data Using the HT, the satellite requires a TOF of 24.10 TU (1 TU = 13.44686457 minutes). The OTB provides a TOF of 23.45 TU to transfer the satellite from NEO to GEO. (See Appendix A for the calculations supporting these results.)

Interpretation The OTB requires less time than the HT to transfer the satellite from NEO to GEO. Because time-sensitive constraints do not exist in this case, and because the differences in TOF for both options are relatively small, no preference exists for either choice.

Required Fuel

Explanation As mentioned earlier, the energy to move the satellite from NEO to GEO is provided by onboard fuel, which also powers the onboard thrusters for station-keeping activities on orbit. Fuel not required for the orbital transfer would then be available for future on-orbit maneuvers.

The satellite moving in NEO possesses considerable kinetic energy, as shown in Equation (1).

$$E = \frac{1}{2}mv^2 \tag{1}$$

While the mass m of the satellite is considered to be constant, the change in velocity v requires an increase in energy which, in turn, requires more fuel.

Data The total change in velocity for the HT is .498 DU/TU (1 DU/TU = 7.90536828 km/s). The total change in velocity for the OTB is .594 DU/TU. (See calculations in Appendix B.)

Interpretation The HT requires a smaller change in velocity, thereby requiring less fuel compared to the OTB. In terms of fuel savings, the HT is preferred.

Conclusion
Summary
This study evaluated both the HT and OTB based on three criteria: accuracy, time of flight, and fuel required. The HT's performance was better in terms of accuracy and required fuel. The OTB's performance was better in terms of TOF (See Table 8.1).

Table 8.1 Summary of results

Maneuver	Accuracy	TOF	Change in v
HT	85.5%	24.10 TU	.498 DU/TU
OTB	60.2%	23.45 TU	.594 DU/TU

Conclusions
The HT is preferred based on accuracy and fuel requirements. The OTB performed better in terms of time of flight, but the advantage is not considered significant given the constraints of the problem.

Recommendation
The HT is recommended as the transfer maneuver for moving the Big Ears satellite from NEO to GEO.

Contact
For more information, contact James R. McFinkel, Ph.D., NIA/OR3, at (202) 444-9999.

Source
Orbitus, Delta V.: *The Joy of Orbital Transfer.* Dayton, Ohio: Astro Press, 2004.

Appendix
Calculations are not included for the purposes of this example.

The following example is a feasibility report on using the 16XL1000000 tube in a new, high-power shortwave transmitter application. The organization follows Outline 8.1. Notice that this type of report differs from a recommendation report in that it evaluates only a single solution. The focus is on applying criteria and interpreting data to determine the feasibility of this single solution, rather than comparing multiple solutions.

Feasibility Report: Using the 16XL1000000 in the *Boss*RF Type 5000 Transmitter

Introduction

PURPOSE

The purpose of this report is to assess the feasibility of using the 16XL1000000 Megatube in the *Boss*RF Type 5000 shortwave transmitter.

PROBLEM

*Boss*RF Systems is retrofitting and upgrading its existing Type 3000 shortwave transmitter to provide a new, high-power, shortwave broadcast transmitter (designated Type 5000). Type 5000 will be capable of continuous, high-level amplitude modulation between 3.0 and 26.1 megahertz with a final input power of 1 megawatt. Type 3000 currently uses the Eimac 4CM400,000A as the RF power amplifier.

This tube will be inadequate for the new power levels. As a replacement, the XLTubes 16XL1000000 is being considered. At this point, an initial feasibility assessment for its use is required. The critical operational considerations for this preliminary assessment include the tube's size, performance, and cost.

SCOPE

This report provides an initial feasibility assessment, and as such, it will evaluate the 16XL1000000 as an RF power amplifier based solely on size compatibility, power ratings, and cost. Other considerations, such as cooling requirements and health and safety issues, will not be addressed in this report.

Discussion

SIZE

Explanation

Size refers to the volume and specific dimensions of the space requirements for the tube. The tube, including its socket, cavity, and associated components, must fit properly in the existing power amplifier cage of the Type 3000 transmitter.

Data

The Type 3000 transmitter's final amplifier cage provides 150,000 cubic centimeters of space for the power amplifier tube and associated electrical and cooling connections, as well as final amplifier tuning inductors and capacitors. The usable space is 50 centimeters deep, 50 centimeters wide, and 60 centimeters high. The 16XL1000000 occupies a volume of 36,000 cubic centimeters, with a diameter of 30 centimeters and a height of 40 centimeters. Fully installed, with associated power amplifier circuitry and cooling connectivity in place, the 16XL1000000 assembly would occupy between 80,000 and 90,000 cubic centimeters with a diameter of 40 centimeters and a height of 50 centimeters.

Interpretation

The 16XL1000000, as operationally configured with associated power amplifier circuitry and cooling connectivity, will fit in the existing power amplifier cage of the Type 3000 transmitter.

PERFORMANCE

Explanation

Performance includes these key operational capabilities: continuous power input of 1 megawatt in high-level, amplitude modulation (A3) service, 70 percent efficiency in the frequency range between 3.0 megahertz and 26.1 megahertz, and full power output with 2000 watts of drive. (*Boss*RF)

Data

The 16XL1000000 is capable of a maximum continuous plate dissipation of 1,200,000 watts with an amplification factor of 6.2. Efficiencies exceed 72 percent up to a maximum useful frequency of 150 megahertz with a 2000-watt input signal. (BossRF)

Interpretation

This tube meets the specified performance requirements.

COST

Explanation

Cost refers to the catalog price from the manufacturer for both new and rebuilt 16XL1000000 tubes. This criterion takes on added importance because a second

16XL1000000 may well be required as a high-level audio modulator, although the scope of this report is limited to the tube's use as an RF power amplifier.

Data

The manufacturer's catalog price for the 16XL1000000 is $129,000 new and $98,000 rebuilt. (XLTubes)

Interpretation

No alternative, single-tube technology to the 16XL1000000 exists. The cost of redesigning the Type 5000 transmitter to use two, smaller 500,000-watt power tubes has not been established. Additionally, the cost of each of these smaller tubes typically approaches 80 percent of the cost of the 16XL1000000 (BigTubes, Inc.). Therefore, the 16XL1000000 is cost-effective.

Conclusion

SUMMARY

The feasibility for the use of the16XL1000000 in the Type 5000 transmitter was evaluated based on size, performance, and cost. In terms of size, the tube will fit in the current Type 3000 transmitter's power amplifier cage assembly. The tube's performance will meet all operational requirements for the Type 5000. The 16XL1000000 is also cost-effective.

CONCLUSIONS

The 16XL1000000 is a feasible power amplifier tube for the upgraded Type 5000 transmitter based on its size, performance, and cost.

RECOMMENDATION

The 16XL1000000 should be considered a viable solution in terms of size, performance, and cost. However, additional studies should be performed to assess the 16XL1000000's feasibility for the Type 5000 in terms of cooling and heat dissipation requirements as well as health and safety considerations.

CONTACT

Frederick D. Phillips, Chief of Analysis, *Boss*RF Systems, extension 5676, e-mail: fdphillips@analysis.bossrf.com.

SOURCES

BigTubes, Inc.: "Big Power and Big Costs," Internet: http://www.bigtubesinc.com, January 2004.
*Boss*RF Systems, 16XL1000000 specification sheet, August 2004.
 XLTubes, "Online Price Lists," Internet: http://www.xltubes.com, January 2004.

Feasibility and Recommendation tion Report Checklist

- Have I defined the problem with the necessary level of detail to ensure the requirements for solving it are clear?
- Have I selected a manageable number of candidate solutions that are apparently viable?
- Have I developed criteria that relate to the problem?
- Have I explained all criteria, including why they were selected and how much weight each is being given?
- For all criteria, have I collected information (data) that is objective and meaningful?
- Have I provided useful interpretations of this information (these data)?
- Have I included a conclusion based on these interpretations?
- Have I made a recommendation based on this conclusion?
- Have I included a contact who can provide more information about this report?
- Have I documented the sources I used for my information?
- Have I included in the appendix any necessary supporting documents?

09
01001

Laboratory and Project Reports

Remember those old horror movies in which the insane scientist transplanted human brains from one body to another in the laboratory? The labs were always impressive. All around one could see and hear high-voltage arcs traveling up Jacob's ladders and corona discharges zapping off spherical electrodes. What those movies did not show was the scientist typing up a laboratory or project report after the experiments were concluded. One does not successfully transplant a brain from one body to another and not write a report!

Virtually anyone working in engineering and the sciences will be tasked, in one fashion or another, with producing a laboratory or project report. These documents present information that relates to the controlled testing of a hypothesis, theory, or device using test equipment (the apparatus) and a specified series of steps employed to perform the test (the procedure). The emphasis in a laboratory or project report is on documenting the design and conduct of the test, how the variables were controlled, and what the resulting data show.

What Are Laboratory and Project Reports?

In their purest form, laboratory reports are research-oriented documents, meaning that they start with a hypothesis or theory that needs to be applied and tested under highly controlled conditions. For example, suppose you are an aeronautical engineer hypothesizing that your new wing design could be used to generate higher lift at hypersonic speeds with increased flight stability. To test that hypothesis in a laboratory, you would need an apparatus—in this case, a hypersonic wind tunnel and a model of your wing design. You would also need a procedure for using that wind tunnel to test your wing design model. You could then use the procedure to collect data from the wind tunnel tests and interpret the data to see if your new wing design generated higher lift with increased stability under hypersonic conditions. Finally, you could assess whether the original hypothesis was supported and, if so, probably recommend that more research be done.

Laboratory reports can also take the form of project reports, which are commonly used in teaching laboratories. Instead of testing a hypothesis or theory, you would focus on fulfilling specific requirements that normally come from your instructor in the form of a project assignment. The goal of the project might be to demonstrate the application of a theory or set of theories by using available technology. For example, you and several other students in an engineering project course might be asked to build a motorized vehicle with a homeostatic control system that would track accurately along a 100-yard course. The course could be marked with a white line and could contain six $90°$ curves. Your group's vehicle also would have to avoid several obstacles placed in its path along the way while completing the course within a specified time.

As students, you would design and build an operating vehicle within the cost and equipment constraints prescribed by the teacher. Your group would probably apply basic control theory, using optical and ultrasonic sensors, several electromagnetic servos, and an embedded microprocessor. In effect, your vehicle and the test track would be the apparatus, and the process used to test your vehicle's performance would be the procedure. As part of the class, your group would make several test runs of the vehicle, collect data relating to speed and accuracy, and then develop a project report to document how well your vehicle met the requirement.

Laboratory reports and project reports are similar and generally follow the same pattern, except for a few minor differences. Outline 9.1 provides a model for a research-oriented laboratory report. Outline 9.2 provides a slightly different model for the kind of project report frequently required as part of a teaching laboratory. In our first example,

---

Outline 9.1 Laboratory Report

Introduction	
• Purpose	Describe the reason for writing this report.
• Problem	Describe the context and hypothesis(es) for this report.
• Scope	Describe the limitations of this report.
Background	
• Theory	Review the theoretical basis of this research.
• Research	Review prior research relevant to this research.
Test and evaluation	
• Apparatus	Describe device(s) used to do this research.
• Procedure	Describe procedure(s) used to do this research.
Findings	
• Data	Review the results of the test.
• Interpretation	Provide your interpretation of the results.
Conclusion	
• Assessment	State whether, and to what extent, the hypothesis is supported.
• Recommendations	Provide your recommendation(s), if any.

Outline 9.1 will be used to develop a research laboratory report on the 16XL1000000 transmitting tube. Following that, Outline 9.2 will be used to produce a project report for an undergraduate civil engineering course.

Outline 9.2 Project Report

Introduction

- Purpose   Describe the reason for writing this report.
- Problem   Describe the context for this report, including project requirements.
- Scope   Describe the limitations of this report.

Background

- Theory   Review the theoretical basis for responding to requirements.
- Research   Review prior research relevant to requirements.

Test and evaluation

- Apparatus   Describe device(s) used to accomplish the task.
- Procedure   Describe procedure(s) used to accomplish the task.

Findings

- Data   Review the results of the test and evaluation.
- Interpretation   Provide your interpretation of the results, that is, to what extent requirements were met.

Conclusion

- Assessment   State your conclusions based on the interpretation(s).
- Recommendations Provide your recommendation(s), if any.

Research Oriented Lab Report

Our first example is also the threaded example. The following illustrates a research-oriented laboratory report. This report documents preliminary laboratory testing on the 16XL1000000 Megatube, assessing the device's performance in audio service.

Introduction

Start the introduction by briefly and succinctly describing the purpose of the report.

Purpose

This report documents the preliminary testing and analysis of the 16XL1000000 Megatube to determine its effectiveness when used as an audio-frequency (AF) power amplifier for rock concerts held in and around large arenas.

Next, provide the reader with enough information on the problem so that he or she can understand what the report is about and put the information that follows in the proper context. For this example, the problem involves a fictitious rock group, the Village Thumpers, whose act depends on high-amplitude, low-frequency "thumping" sound waves to ensure a truly measurable impact on the audience.

Problem

The Village Thumpers, a rock group known for its high-amplitude, megabass music, is planning an extended concert series in a large stadium (> 80,000 seats). The audio system must produce low-frequency pressure waves with amplitudes capable of disorienting audience members through vestibular displacement anywhere within the stadium or its immediate surrounding area (approximately 20 acres). To meet the low-frequency audio requirement, the sound system will use the enclosed stadium as a resonant baffle, with four arrays at each corner containing 25 woofers rated at 1 kilowatt each. With a cumulative power requirement for the four arrays of 100 kilowatts, and assuming a nominal loss of 10 decibels, the amplifier must produce at least 1 million watts of audio to provide the necessary drive to the four woofer arrays.

Solid-state amplifiers are not as feasible as vacuum tubes at this power level. This power level would require hundreds of thousands of power transistors operating in parallel in small groups through the use of cascaded combiner transformers to equal the output of a single vacuum tube. Additionally, electron tube amplifiers are being used increasingly as power amplifiers in demanding audio environments because they may provide more linear sound reproduction with a richer blend of harmonics. It is therefore hypothesized that the 16XL1000000 will meet the audio requirements of the Village Thumpers stadium concert.

Round out the introduction with a statement of scope regarding what is being tested and any significant limitations of these tests.

Scope

This preliminary test and evaluation of the 16XL1000000 in a power audio amplifier application includes only an analysis of power, distortion, and noise. Furthermore, this report is limited to the technical performance of the device and does not include analysis of its cost-benefit ratio or marketability in audio applications.

Background

In this section, provide the information necessary for the reader to understand and appreciate the test report and the findings that will follow. This section should review any relevant theory and past research that the reader needs to know.

Theory

Vacuum tubes were first developed in the early 1900s to control the flow of electronic current and were used virtually in all audio amplifier equipment until the early 1960s. At that time, solid-state devices quickly replaced tubes in audio applications because of their smaller size and higher efficiencies, except in very high-power applications where solid-state devices were not as feasible.

Additionally, in recent years there has been a noted resurgence in audio amplifier applications using vacuum tubes for reasons other than power. Apocryphal data indicate that many audiophiles, audio engineers, and professional musicians prefer the tube amplifier's "warm and softer" sounds. Some attribute the tube's comparatively better sound, vis-à-vis solid state, to some of or all the following: better power management, less distortion, improved signal-to-noise ratios, better linearity, less clipping, and fewer feedback problems. This preliminary analysis focuses on the first three variables—power, distortion, and noise.

The 16XL1000000 is a 1.2-megawatt transmitting tube designed primarily as a wideband, radio-frequency (RF) power amplifier running in class C mode across the spectrum from .5 kilohertz to 300 megahertz. This laboratory research is designed to test the performance of the 16XL1000000 in audio-frequency (AF) service, running as a power amplifier in class A across the spectrum from direct current (dc) to 20 kilohertz.

Research

Limited research has been done on using RF power tubes in AF applications, although Williams did investigate the limited use of low-power transmitting tubes in audio applications. (Williams 2000, pp. 347–350.) This investigation indicated that while the tube could function effectively at audio frequencies and provide solid performance at its rated RF output power, the cost factors were not favorable for such applications. This analysis predates the current resurgence in tube amplifier applications. It also did not address high-power applications that lend themselves to the state-of-the-art technology capabilities of the 16XL1000000.

Test and Evaluation

In this section, describe both the physical devices or *apparatus* used in the test and the processes or *procedure* for doing the testing. The apparatus includes the device being tested and the equipment used to do the testing, while the procedure would include the steps in the test.

Apparatus

The 16XL1000000 is capable of producing a continuous output of 1.2 million watts of audio into large arrays with an input signal of 1000 volts peak to peak. This power is more than adequate for demanding environments such as rock concerts, and it will even meet the controversial pressure wave requirements of the Village Thumpers—that is, disorienting audience members through vestibular displacement anywhere within the stadium or its immediate surrounding area.

Local zoning ordinances and environmental protection considerations would not permit laboratory evaluation of the 16XL1000000 at its maximum rating. Consequently, for test purposes, the 16XL1000000 was used in a single-ended audio amplifier, running in class A. The tube was driven with a multi-stage, high-gain voltage amplifier with a nominal output power of 100 watts.

The voltage and power amplifiers were combined into the amplifier test unit (ATU). The ATU was constructed on a steel chassis with proper external and interstage shielding, and the power amplifier stage employed appropriate cavity and coaxial socket interfaces for the 16XL1000000. A calibrated signal generator was used to drive the amplifier. The output was fed into a purely resistive 8-ohm load, where it was analyzed by wide-bandwidth, multitrace oscilloscopes as well as high-end computers running proprietary analysis software.

Procedure

A pure, distortion-free sine wave signal of 1 volt peak to peak was applied to the voltage amplifier stages of the ATU. This level was the maximum input signal amplitude possible without saturating these stages. The frequency of the signal was varied across a range from dc to 20 kilohertz in 1-kilohertz increments. The output was analyzed in terms of power management, distortion, and noise.

Findings

In this section, present the data that the tests yielded and provide an interpretation of these data. As you will see later, a table and various graphs can be used to display the data.

Data

The data gathered fell into these three categories: power in watts, distortion as a percentage of total signal, and signal-to-noise ratio.

Power

Power is the product of voltage and current output from the amplifier. Constancy of power across the full frequency range is essential for properly driving complex speaker systems in demanding environments. The power output of the 16XL1000000 was observed across the full range of frequencies tested. Although the output varied slightly across the test range and clearly dropped off above 18 kilohertz, overall the output remained relatively constant. This constancy was especially true in the primary audio frequencies between dc and 15 kilohertz.

Distortion

Distortion is the unwanted characteristic of an amplifier to modify the nature of a signal while amplifying it. Although solid-state amplifiers are often credited with having lower distortion values than tube amplifiers, the 16XL1000000 exhibited harmonic distortion levels across its total operating range of less than .05, which is as good as, if not better than, most solid-state amplifiers. The distortion that did occur was at the upper limit of the range, well above the threshold of human hearing.

Noise

Noise includes spurious output signals that were not part of the original input signal. Good solid-state amplifiers typically have a signal-to-noise ratio (SN) of .15. By comparison, the 16XL1000000 provided a typical SN of .09 to .15 across the primary audio spectrum.

Interpretation

In terms of the preliminary tests involving power, distortion, and noise, the 16XL1000000 performed well in an audio-frequency power amplifier mode. Although the high-power levels could not be duplicated in the laboratory setting, the low-power tests demonstrated excellent results, as the following data show.

Data:

khz	P	dis	SN	khz	P	dis	SN
1	10.22	.00	.15	11	10.31	.02	.10
2	10.24	.00	.15	12	10.30	.01	.11
3	10.26	.01	.10	13	10.31	.02	.11
4	10.30	.01	.10	14	10.33	.01	.11
5	10.29	.01	.09	15	10.21	.01	.12
6	10.29	.01	.09	16	10.30	.03	.12
7	10.30	.01	.09	17	10.29	.03	.13
8	10.30	.02	.09	18	10.00	.03	.14
9	10.29	.09	.10	19	10.00	.03	.14
10	10.29	.01	.10	20	10.00	.04	.15

Here khz = frequency in kilohertz; P = power in watts; dis = distortion in percent; SN = signal-to-noise ratio.

Conclusion

Finally, provide your overall conclusions related to the original purpose of this study, and make any specific recommendations that you believe are warranted by the results.

ASSESSMENT

Initial laboratory analysis, based on the 16XL1000000's measured performance, supports the hypothesis; the tube seems to be a viable candidate for the sound system supporting the Village Thumpers stadium concert.

RECOMMENDATION

The Village Thumpers should consider the 16XL1000000 a technically suitable option for audio power amplifier applications.

Source

Williams, Robert E.: "Investigation of Transmitting Tubes for Audio Applications," *Technical Tidbits*. New York: SoundProducts International, 2000, pp. 345–354.

Here is the laboratory report with sources, charts, and table included.

Putting It All Together

Preliminary Report on the 16XL1000000 in Audio Service

Introduction
Purpose

This report documents the preliminary testing and analysis of the 16XL1000000 transmitting tube to determine its effectiveness when used as an audio-frequency power amplifier for rock concerts held in and around large arenas.

Problem

The Village Thumpers, a rock group known for its high-amplitude, megabass music, is planning an extended concert series in a large stadium ($> 80,000$ seats). The audio system must produce low-frequency pressure waves with amplitudes capable of disorienting audience members through vestibular displacement anywhere within the stadium or its immediate surrounding area (approximately 20 acres). To meet the low-frequency audio requirement, the sound system will use the enclosed stadium as a resonant baffle, with four arrays at each corner containing 25 woofers rated at 1 kilowatt each. With a cumulative power requirement for the four arrays of 100 kilowatts, and assuming a nominal loss of 10 decibels, the amplifier must produce at least 1 million watts of audio to provide the necessary drive to the four woofer arrays.

Solid-state amplifiers are not as feasible as vacuum tubes at this power level. This power level would require hundreds of thousands of power transistors operating in parallel in small groups through the use of cascaded combiner transformers to equal the output of a single vacuum tube. Additionally, electron tube amplifiers are being used increasingly as power amplifiers in demanding audio environments because they may provide more linear sound reproduction with a richer blend of harmonics. It is therefore hypothesized that the 16XL1000000 will meet the audio requirements of the Village Thumpers stadium concert.

Scope

This preliminary test and evaluation of the 16XL1000000 in a power audio amplifier application includes only an analysis of power, distortion, and noise. Furthermore, this report is limited to the technical performance of the device and does not include analysis of its cost-benefit ratio or marketability in audio applications.

Background

Theory

Vacuum tubes were first developed in the early 1900s to control the flow of electronic current and were used in virtually all audio amplifier equipment until the early 1960s. At that time, solid-state devices quickly replaced tubes in audio applications because of their smaller size and higher efficiencies, except in very high-power applications where solid-state devices were not as feasible.

Additionally, in recent years there has been a noted resurgence in audio amplifier applications using vacuum tubes for reasons other than power. Apocryphal data indicate that many audiophiles, audio engineers, and professional musicians prefer the tube amplifier's "warm and softer" sounds. Some attribute the tube's comparatively better sound, vis-à-vis solid state, to some of or all the following: better power management, less distortion, improved signal-to-noise ratios, better linearity, less clipping, and fewer feedback problems. This preliminary analysis focuses on the first three variables—power, distortion, and noise.

The 16XL1000000 is a 1.2-megawatt transmitting tube designed primarily as a wideband, radio-frequency (RF) power amplifier running in class C mode across the spectrum of .5 kilohertz to 300 megahertz. This laboratory research is designed to test the performance of the 16XL1000000 in audio-frequency (AF) service, running as a power amplifier in class A across the spectrum from dc to 20 kilohertz.

Research

Limited research has been done on using RF power tubes in AF applications, although Williams did investigate the limited use of low-power transmitting tubes in audio applications. (Williams 2000, pp. 347–350.) This investigation indicated that while the tube could function effectively at audio frequencies and provide solid performance at its rated RF output power, the cost factors were not favorable for such applications. This analysis predates the current resurgence in tube amplifier applications. It also did not address high-power applications that lend themselves to the state-of-the-art technology capabilities of the 16XL1000000.

Test and Evaluation

Apparatus

The 16XL1000000 is capable of producing a continuous output of 1.2 million watts of audio into large arrays with an input signal of 1000 volts peak to peak. This power is more than adequate for demanding environments such as rock concerts; and it will even meet the controversial pressure-wave requirements of the Village Thumpers, that is, disorienting audience members through vestibular displacement anywhere within the stadium or its immediate surrounding area.

Local zoning ordinances and environmental protection considerations would not permit laboratory evaluation of the 16XL1000000 at its maximum rating. Consequently, for test purposes, the 16XL1000000 was used in a single-ended audio amplifier, running in class A. The tube was driven with a multistage, high-gain voltage amplifier with a nominal output power of 100 watts.

The voltage and power amplifiers were combined into the amplifier test unit (ATU). The ATU was constructed on a steel chassis with proper external and interstage shielding, and the power amplifier stage employed appropriate cavity and coaxial socket interfaces for the 16XL1000000. A calibrated signal generator was used to drive the amplifier. The output was fed into a purely resistive 8-ohm load, where it was analyzed by wide-bandwidth, multitrace oscilloscopes, as well as high-end computers running proprietary analysis software.

Procedure

A pure, distortion-free, sine wave signal of 1 volt peak to peak was applied to the voltage amplifier stages of the ATU. This level was the maximum input signal amplitude possible without saturating these stages. The frequency of the signal was varied across a range from dc to 20 kilohertz in

1-kilohertz increments. The output was analyzed in terms of power management, distortion, and noise.

Findings

Data

The data gathered fell into these three categories: power in watts, distortion as a percentage of total signal, and signal-to-noise ratio.

Power   Power is the product of voltage and current output from the amplifier. Constancy of power across the full frequency range is essential for properly driving complex speaker systems in demanding environments. The power output of the 16XL1000000 was observed across the full range of frequencies tested. Although the output varied slightly across the test range and clearly dropped off above 18 kilohertz, overall the output remained relatively constant. This constancy was especially true in the primary audio frequencies between dc and 15 kilohertz. See Table 9.1 and Figure 9.1.

Table 9.1   Test Data

kHz	P	dis	SN	kHz	P	dis	SN
1	10.22	0.00	0.15	11	10.31	0.02	0.10
2	10.24	0.00	0.15	12	10.30	0.01	0.11
3	10.26	0.01	0.10	13	10.31	0.02	0.11
4	10.30	0.01	0.10	14	10.33	0.01	0.11
5	10.29	0.01	0.09	15	10.21	0.01	0.12
6	10.29	0.01	0.09	16	10.30	0.03	0.12
7	10.30	0.01	0.09	17	10.29	0.03	0.13
8	10.30	0.02	0.09	18	10.00	0.03	0.14
9	10.29	0.02	0.10	19	10.00	0.03	0.14
10	10.29	0.01	0.10	20	10.00	0.04	0.15

Here kHz = frequency in kilohertz; P = power in watts; dis = distortion in percent; SN = signal-to-noise ratio.

Distortion   Distortion is the unwanted characteristic of an amplifier to modify the nature of a signal while amplifying it. Although solid-state amplifiers are often credited with having lower distortion values than tube amplifiers, the 16XL1000000

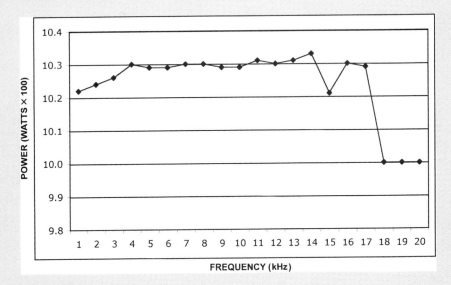

Figure 9.1
Power and frequency.

exhibited harmonic distortion levels across its total operating range of less than .05, which is as good as, if not better than, most solid-state amplifiers. The distortion that did occur was at the upper limit of the range, well above the threshold of human hearing. See Table 9.1 and Figure 9.2.

Noise Noise includes spurious output signals that were not part of the original input signal. Good solid-state amplifiers typically have a signal-to-noise ratio (SN) of .15. By comparison, the 16XL1000000 provided a typical SN of .09 to .15 across the primary audio spectrum. See Table 9.1 and Figure 9.3.

Interpretation

In terms of the preliminary tests involving power, distortion, and noise, the 16XL1000000 performed well in an audio-frequency power amplifier mode. Although the high-power levels could

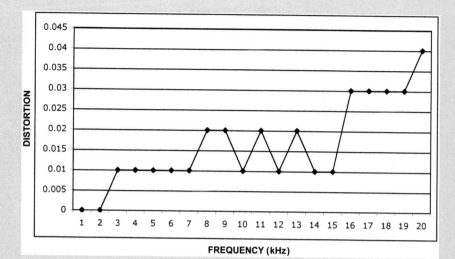

Figure 9.2
Distortion and frequency.

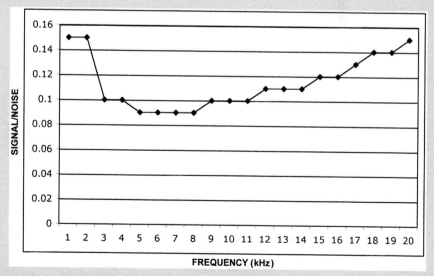

Figure 9.3
Signal/noise ratio and frequency.

not be duplicated in the laboratory setting, the low-power tests demonstrated excellent results (Table 9.1).

Conclusion
Assessment
Initial laboratory analysis, based on the 16XL1000000's measured performance, supports the hypothesis; the tube seems to be a viable candidate for the sound system supporting the Village Thumpers stadium concert.

Recommendation
The Village Thumpers should consider the 16XL1000000 a technically suitable option for audio power amplifier applications.

Source
Williams, Robert E.: "Investigation of Transmitting Tubes for Audio Applications," *Technical Tidbits*. New York: SoundProducts International, 2000, pp. 345–354.

Student Project Report

The following is a hypothetical example of a short project report based on Outline 9.2. The report is required by a fictitious civil engineering course where small groups of students are tasked to design a system for measuring the average height and slope of a road surface. The height and slope data are necessary for repaving an existing asphalt road.

Acquiring Height and Slope Data for Paving Operations

Introduction

Purpose
This report documents the analysis required for the course project assigned to group 2 to fulfill in part the requirements of CE 400, Civil Engineering: Technology and Methods.

Problem

Our group consists of four seniors majoring in civil engineering. The group was assigned the task of designing a system that would permit road pavers to obtain average height and slope data of a 70-foot segment of two-lane road. The requirements included being able to supply data to pavers in "real time" to allow them to lay down an asphalt mat that is precise in thickness and smoothness and that fills in surface irregularities, whether severe or gradual in nature. The system must be constructed from off-the-shelf components, be electronic in nature and operation, and be designed and fabricated within 7 weeks at a total cost not to exceed $200 (see course tasking, Appendix A).

Scope

This report documents the approach, test, and evaluation of group 2's laboratory project for CE 400, Fall 2005. We noted that the course tasking only requires demonstration that the data generated by the system could be used for paving operations. We assumed, therefore, that no requirement exists to test our solution in actual paving operations.

Background

Research

Paving an asphalt road with precise mat thickness and smoothness requires an averaging system that can track variations in the height and slope of a road's surface, and provide this information in real time to construction equipment conducting paving operations. Traditionally, obtaining this kind of data requires either a mechanical or electronic approach. Mechanical systems typically use grade-leveling devices similar to "skis" that extend in front of and behind the paver. The front ski senses irregularities in the road, while the trailing ski provides data on the smooth surface that has just been paved. These skis provide the average height and slope for the road, which are then matched in real time by the *screed,* which controls the height and slope of the asphalt mix being applied. Mechanical skis can be effective, but they are easily damaged by road hazards, require high labor costs to set up, and wear rapidly because of friction with the asphalt surface (Koko 2003, pp. 58–60).

Electronic approaches to providing paving data use emitted radio-frequency, acoustic, or light energy that reflects off the road's surface and is then sensed and read over time and distance to map this surface. Often called *time-of-flight* systems, these devices map the road's surface by measuring the time required by the emitted energy to travel to the surface and back. Several approaches using emitted energy exist, including (but not limited to) ultrasonic, laser, pulsed laser, active triangulation, and continuous RF wave that is either amplitude- or frequency-modulated (Tyger 2005, p. 34).

To respond to the course tasking criteria, especially those placing severe constraints on time and cost, our group chose the ultrasonic solution. Ultrasonic transducers are widely available and relatively inexpensive, and the apparatus required to function at acoustic frequencies is easier to fabricate than those operating in optical or radio-frequency spectra. Typically, off-the-shelf ultrasonic range finders costing less than $150 are capable of measuring distances up to 30 feet (Peesher 2003, p. 136).

Theory
Ultrasonic range finders usually operate using a pulsed method, where a ping is emitted and its echo is measured. The sound waves of the ping propagate from the transducer, bounce off the road's surface, and echo back to the receiver. Since the speed of sound can be determined accurately based on air temperature, the time between ping and echo can be converted to precise distance measurements. The constant mapping of distance from a moving vehicle can provide an accurate sampling of irregularities in, and slope of, the road's surface over which the vehicle is moving.

To determine the average height of the road, several tasks must be accomplished: (1) Select a cartesian coordinate system with a precisely defined origin. (2) Create a system of xy plots from the perspective of the origin by transforming a synchronized set of sensor-to-road measurements. (3) Calculate the best-fit line between these points. (4) Adjust the application of asphalt by the paver to match this line. Our group developed an algorithm to accomplish tasks 1 through 3, a limitation that is consistent with the scope of this project (Appendix B).

Test and Evaluation

The group's approach was to construct an ultrasonic range finder that could be mounted on a moving vehicle. This approach employed a computerized recording device to track variations in distance, process this information based on our averaging algorithm, and produce a physical profile of the road's surface that could be used to control the application of asphalt.

Apparatus
Our group acquired, at a local flea market for $100, a Poladucer 500 transducer commonly used in consumer devices. We designed a range-finding circuit (Appendix C) that samples with a controlled, narrow beam directed at a 90° angle to the road's surface. Temperature sensors mounted at the transducer and near the road's

surface provided the thermal gradient data necessary to properly estimate the propagation delay of the acoustic energy. We used a Pentium IV notebook computer "borrowed" from an unattended electrical engineering lab to do the necessary I/O functions and processing, employing interfaces and code produced by our group (Appendix D). A 2.1-megapixel digital camera was used to create photographic sequences of the road's surface.

Procedure

To test our surface-measuring device, we mounted it on the trailer hitch of a pickup truck. The sensor looked straight down at a right angle to the road 2.5 feet below. The computer was operated in the cargo bed of the truck.

To sample a smooth strip of road for test purposes, we drove the truck down the well-kept driveway to the president's house on the university campus. To gather data from a rough road surface, we conducted the same test on the decaying driveway leading to the on-campus student apartments. For both tests, we stopped the truck every 10 feet and physically measured the distance to the road's surface. We flagged the data taken at that point and compared it with the physical measurements. We also produced a cross-sectional profile of the road's surface and visually compared that computer-generated

surface plot to sequenced photographs of the actual road surface (Appendix E).

Findings

Data

The test data are provided in Appendix F, along with the profiles taken of the road surfaces. Generally, the data tracked well with the physical measurements, with accuracy of .2 to 1.5 percent of the total range. Additionally, the profiles constructed closely matched the visual appearance of the road's surface. However, we did note a substantial reduction in accuracy (2.5 to 5 percent of total range) where surface features changed suddenly. For example, our system was not able to handle the sharp vertical offsets associated with potholes encountered on the road to the student housing area.

Interpretation

We believe our system meets the assignment criteria of the course. Arrays of overlapping ultrasonic transducers could be fitted onto traditional paving vehicles, thereby replacing the front and rear mechanical skis. A computer onboard the paver could be programmed to sample all the devices in the arrays. Together, these devices would provide a much wider and higher-resolution view of the road than the single, narrow strip our test apparatus sampled. This improved view would enhance the capability to deal with sudden changes or irregularities in the road's surface, and would enable better control for the application of asphalt in paving operations.

Conclusion

Assessment
We assess the performance of our group as *excellent*. We have developed a simple yet effective electronic device that can measure the height and slope of a road's surface and translate these measurements to control signals from a computer, to manage in real time the application of asphalt from a paver.

Recommendations
For obvious reasons involving cost and safety, our group could not obtain the services of a paver, construction access to a deteriorated road that needed to be paved, or the expertise and materials necessary to actually pave it. We recommend that the university adapt our device for an operational paver and conduct a true field test by repaving the driveway leading to student housing.

References

Siamesus F. Koko, "The Problem with Mechanical Skis," *Paving Quarterly,* (2003), pp. 50–65.

Elmer S. Tyger, "Time of Flight Can Do It Right," *Paving Quarterly* (2005), pp. 30–38.

Stephen B. Peesher, "Range Finding Distances: The Efficacy of Evolving RF Systems," 22nd ISMEAX Conference, Anchorage, Alaska, June 2003, pp. 131–159.

Laboratory and Project Report Checklist

- Have I clearly defined the purpose of this report?
- Have I clearly described the problem that requires this report?
- Have I clearly explained the limitations of this report?
- Have I discussed any theory necessary for the reader to understand the report?
- Have I reviewed relevant prior research?
- Have I described the apparatus I used to collect the data?
- Have I described the procedure I used to collect the data?

- Have I clearly provided the results of the test?
- Have I properly interpreted these results?
- Have I made proper conclusions from these interpretations?
- Have I made a recommendation based on my conclusions?

10
01010

Instructions and Manuals

Explaining to someone how to do something can be a challenging task. If you do not believe that, try showing a young child how to tie his or her shoes. Both instructions and manuals describe processes so that readers can accomplish the steps required to successfully complete a task. In technical writing, instructions deal with narrowly defined topics where the goal is to explain how to do something and not much more. Broader requirements demand complex descriptions in the form of manuals. Both instructions and manuals have the same objective, but manuals contain more than just instructions for carrying out a simple task. Fundamentally, however, they both do the same thing and work the same way.

What Are Instructions?

Instructions are process descriptions for human involvement. You probably remember that process descriptions were discussed in Chapter 5. The process description is a good starting point for writing instructions, but instructions involve more than simply describing a process. When you give instructions, you are describing not only the steps of a process, but also how to accomplish these steps. That means you are accepting the additional responsibilities of describing the process accurately and in a way that your reader can follow safely and effectively. It is the difference between describing to a medical student what happens when an inflamed appendix is removed and describing to a surgical resident, hovering over a patient with scalpel in hand, how to remove an inflamed appendix. To appreciate the difference, just imagine that you are the patient!

As Outline 10.1 shows, providing technical instructions is actually straightforward if you understand the process as well as the specific needs and skill level of your audience.

Outline 10.1 Instructions

Introduction
- Definition   Define the overall process.
- Overview   Describe its purpose and provide an overview.
- Theory   Explain any needed theory or principles.
- Steps   List the steps.

Discussion
For each step listed above:
- Definition   Define the step.
- Overview   Describe what happens in this step.
- Background   Provide needed context specific to this step.
- Dangers/cautions   Note any dangers and cautions the reader should be aware of.
- Equipment   List the equipment and tools required for this step.
- Directions   Provide specific directions for executing this step.
- Result   Describe the result that should occur.
- Transition   Move coherently from this step to the next step.

Conclusion
- Summary   Briefly summarize the steps of the process.
- Information   Tell the reader where to find any needed, additional information.

As you might imagine, since the goal of instructions is to show a reader how to actually do something, the reader's skill level is a paramount concern. How you approach instructions is determined substantially by the task at hand and the skill level of the audience. This chapter illustrates how to use Outline 10.1 by developing instructions designed to show a nontechnical audience how to set up an ergonomically correct computer workstation. The second example provides instructions to an expert audience for neutralizing the 16XL1000000 Megatube in a high-power shortwave transmitter—a task no sane layperson would ever attempt.

Instructions for the Layperson

The following example develops a set of general instructions geared to the lay reader.[1] Notice how the theoretical discussions are limited and the language is more appropriate for a nontechnical audience.

Disclaimer

Before we get started, it is worth noting that instructions often carry disclaimers because of potential risks and liabilities. You cannot buy a TV antenna without receiving inserts and brochures warning of the dangers of being hit by lightning, sticking the antenna into nearby power lines, or falling off the roof. So before getting into the specifics of setting up a computer workstation, we might want to add a short disclaimer.

> The following instructions for setting up a computer workstation summarize several ergonomic guidelines for the average person with a typical use profile. However, not all guidelines are included, and no single workstation arrangement can possibly accommodate everyone's ergonomic requirements. In addition, workstations designed for specialized, nonstandard functions may require unique configurations developed by a qualified ergonomics engineer.

Introduction

Following Outline 10.1, begin with a definition of the process, describe the purpose, and then provide an overview of what happens:

Setting up a computer workstation is an ergonomics-based process designed to enhance the user's health, comfort, and productivity. Generally speaking, setting up a computer workstation involves placing and adjusting the desk, chair, and lighting.

Next, discuss any theory or principles *needed* by the reader to understand the setup process. In this case we could address such topics as desk heights and surface glare, dynamic versus static sitting, lower back curvature support, and light intensity and color temperature. We could get into keyboard ergonomics and shoulder, elbow, forearm, wrist, and hand pathology. We also could look at repetitious static work habits and fatigue. But why? Is this information necessary for a layperson who just needs to set up a workstation? For the unskilled audience and the purpose at hand, a discussion of theory and principles is not required. We only need to mention that the process is grounded in ergonomic principles and then simply move on and list the steps.

The setup process follows well-established ergonomic principles and includes the following steps: placing the desk, adjusting the chair, laying out the work area, and optimizing the lighting.

Discussion

At this point, we need to provide specific instructions for each step. Start with step 1 by first defining the step. Then make an overview of what happens in the step and provide necessary information such as danger and caution notices and any needed equipment.

Step 1: Placing the Desk

Placing the desk is the physical action of locating and orienting the primary work surface in the work area. Normally, this step involves moving and orienting the desk that you will be using. The goal is to ensure an effective working location relative to traffic patterns, external light sources, power and network connectivity, and other required equipment. The most important considerations at this time are the location of windows and primary traffic patterns in the workplace, since changing these aspects after the workstation is set up may be difficult. You need to consider the angle and intensity of external light coming through the window (lighting will be addressed in greater detail in step 4), and you need to ensure an effective work area that can be traversed easily by those working in it. In other words, place your desk in a location where the light will not impair your work, a location that enables you and others to move freely.

Caution: Have adequate help on hand and use proper lifting techniques to protect those doing the work, the desk, and the floor beneath.

Equipment: A cargo dolly may be required, depending on the weight and size of the desk and how far it needs to be moved.

Next, provide the specific execution instructions, describe the result, and transition to the next step. Notice the instructions are written in snappy, imperative style where the subject *you* is understood. Also, notice that a brief explanation is provided as to why the action is needed since, given the audience and purpose, we did not provide a preliminary theoretical discussion.

- Place the desk in the desired position. If you are locating the desk by an exterior window, orient the desk at a 90° angle to the window. That way, light from outside will come from the side, not from behind or in front where it would be more likely to cause brightness and glare problems.
- Make sure that the desk is located where adequate access to power, phone, and network connections exists, and where you and others can still move unhindered throughout the work area.

Now transition to the next step.

Once the desk is in its final position with all power, phone, and network connections accessible, you can adjust the chair.

Begin step 2 by first defining the step. Then provide an overview of what happens in the step, along with necessary information and any needed equipment. Notice that, as before, a brief explanation is provided as to why this action is needed.

Step 2: Adjusting the Chair

Adjusting the chair is the physical action of optimizing seat, backrest, and arm positions. Sitting for long periods can be particularly hard on your back, legs, and feet and can impede blood circulation. So you should optimize your chair's adjustments no matter what your size and shape. This step involves setting the seat height for proper forearm position, taking further actions to ensure the feet are resting firmly on the floor (or a footrest), and adjusting the backrest angle to provide proper back support.

Equipment: A chair with height and back support adjustments is a must for properly setting up a computer workstation. Most adjustable chairs do not require tools and can be adjusted with hand-operated knobs and levers.

Next, provide the specific execution instructions, describe the result, and transition to the next step. Again, notice the instructions are written in snappy, imperative style where the subject *you* is understood.

- Sit in the chair in front of the desk in your normal operating position. Make sure you and your chair fit easily under the desk. If not, do an initial height adjustment of the chair. Then check your arm position with your hands on the desk at the normal keyboarding position. Your forearms should be parallel to the floor with an elbow bend of 90°. Change the height adjustment so that your forearms are in the correct position. If you and your chair can no longer fit under the desk, you need a different desk or chair.

- Ensure your feet rest firmly on the floor when your knees are at a 90° to 110° angle. If you need a footrest, get one, but do not leave your feet dangling unsupported above the floor.
- Set the backrest so that your lower back fits snugly against the back cushion, and you are either sitting upright or slightly reclining. Angle the bottom seat cushion so that your thighs are parallel to the floor with your knees and hips at about the same level.

Finally, transition to the next step.

Once the chair is adjusted, you should lay out the work area.

Deal with step 3 in the same way you dealt with the first two steps. Define the step and then provide an overview of what happens, along with necessary information and any needed equipment.

Step 3: Laying Out the Work Area

Laying out the work area is the physical action of optimizing the location and position of the monitor, keyboard, pointing device, and other equipment. Improperly placed monitors can cause you to bend your neck backward to read, which in turn can lead to both neck and back disorders. Also, a large percentage of health problems associated with computer use relate to the keyboard and pointing device. Extended work on computers can cause repetitive stress injuries to the shoulders, elbows, forearms, wrists, and hands. Properly positioning and angling the keyboard and pointing device can help prevent these problems.

Equipment: A monitor, keyboard with adjustable angles, and pointing device. Wrist rests also may be needed for the keyboard and pointing device.

Next, provide the specific execution instructions, describe the result, and transition to the next step.

- While seated at your normal work position, adjust the computer monitor so that you are the correct distance from the screen for reading with your neck in a relaxed position. Place your monitor so that the screen is 18 to 24 inches from your face. While you are sitting upright in a normal work position, if you can reach out and just touch the screen, your monitor is positioned correctly.
- Adjust the monitor's height and position so that you are looking down at a slight angle when reading the screen. For the typical monitor, the top of the screen should be at about your eye level. If you wear bifocals or progressive no-line glasses, you may need to adjust the screen level even lower so that you are not tilting your head back when reading.
- Adjust the keyboard and mouse. Place the keyboard so that, when you are keyboarding, your wrists are straight and your forearms are still parallel to the floor. Adding negative tilt (raising the front edge of the keyboard) or using a keyboard tray under the top surface of the desk may be necessary.
- Position your mouse or other pointing device so that it is within easy reach and at the same height as the keyboard. That way, you will not have to twist your body or lean your shoulder to use it.
- Position wrist rests in front of both the keyboard and the pointing device, but use them only when you are resting your hands and not actually keying or moving the mouse.

Now provide a transition to the next step.

Once you have laid out your work area, you should adjust the lighting.

Deal with step 4 in the same way you dealt with the other steps. Define the step, explain why the step is needed, and then provide an overview of what happens, along with necessary information.

Step 4: Adjusting the Lighting

Adjusting the lighting is the physical action of optimizing the location, angle, and intensity of both exterior and interior illumination sources. One of the most significant problems with computer workstations is visual fatigue, blurred vision, and headaches caused by glare and improper illumination.

External illumination sources tend to vary significantly in intensity and angle. In step 1, you already placed your desk at a 90° angle to any external window to substantially reduce glare problems from external light. However, you may still need to install blinds, shades, or curtains to block out external light sources.

Internal light sources in offices are usually too bright for computer workstation use, with overhead lighting being the single biggest problem. One solution is to adjust the monitor's angle slightly to reduce the glare. Another is to reduce the intensity of the overhead lights, by removing some of the bulbs, installing dimmer circuits where appropriate, or simply switching off banks of lights altogether. The contrast of the display can be increased, and glare screens such as neutral density and polarizing filters can be added in front of the monitor. Neutral density filters reduce the intensity of the glare relative to the intensity of the displayed information, while polarizing filters block the polarity of the reflected glare while permitting the displayed information to pass through. Monitor shields can be installed around the screen to block the source of the glare.

Equipment: It varies significantly with the amount of glare and the solutions required.

Next, provide the specific execution instructions, describe the result, and clarify that the process is complete.

- While you are seated at your normal work position, look at your computer monitor. If you see reflections, you have glare. First, try to remove it by slightly tilting the monitor. If that does not work, identify the source of the glare. Doing so is easy because you can see it reflecting in your monitor's screen.

- If an external window is the problem, you can reduce the glare with shades, blinds, or curtains.
- If internal lights are the problem, move the lights or reduce their intensity. In some cases, you can just turn them off. Use supplemental desk lighting to compensate for the loss of overhead lighting. If that does not work, install glare screens and monitor shields.

When you are able to sit in your normal work position and not see annoying reflections in your monitor's screen, the workstation's setup is complete.

Conclusion

Finally, conclude the instructions by briefly summarizing the steps and providing the reader with sources for additional information.

Setting up a computer workstation is an ergonomics-based process designed to enhance the user's health, comfort, and productivity. Generally speaking, setting up a computer workstation involves optimizing the desk's location, adjusting the chair, arranging the work area, and adjusting the lighting.

For more detailed information on setting up your computer, see "Computer Workstation Ergonomics," Center for Disease Control, Internet: http://www.cdc.gov/od/ohs/ergonomics/compergo.htm.

Putting It All Together Here is the complete instruction set for setting up a workstation, including fully integrated visuals. Note the use of layout and font formatting, as well as white space, to enhance the presentation of information.

Setting Up a Computer Workstation

Disclaimer
The following instructions for setting up a computer workstation summarize several ergonomic guidelines for the average person with a typical use profile. However, not all guidelines are included, and no single workstation arrangement can possibly accommodate everyone's ergonomic requirements. In addition, workstations designed for specialized, nonstandard functions may require unique approaches developed by a qualified ergonomics engineer.

Introduction
Setting up a computer workstation is an ergonomics-based process designed to enhance the user's health, comfort, and productivity. Generally speaking, setting up a computer workstation involves placing and adjusting the desk, chair, and lighting.

The setup process follows well-established ergonomic principles and includes the following steps: placing the desk, adjusting the chair, laying out the work area, and optimizing the lighting.

Discussion

Step 1: Placing the Desk
Placing the desk is the physical action of locating and orienting the primary work surface in the work area. Normally, this step involves moving and orienting the desk that you will be using. The goal is to ensure an effective working location relative to traffic patterns, external light sources, power and network connectivity, and other required equipment. The most important considerations at this time are the location of windows

and the primary traffic patterns in the workplace, since changing these aspects after the workstation is set up may be difficult. You need to consider the angle and intensity of external light coming through the window (lighting will be addressed in greater detail in step 4), and you need to ensure an effective work area that can be traversed easily by those working in it. In other words, place your desk in a location where the exterior light will not impair your work, a location that enables you and others to move freely.

Caution:
Have adequate help on hand, and use proper lifting techniques to protect those doing the work, the desk, and the floor beneath.

Equipment: A cargo dolly may be required, depending on the weight and size of the desk and how far it needs to be moved.

- Place the desk in the desired position. If you are locating the desk by an exterior window, orient the desk at a 90° angle to the window. That way, light from outside will come from the side, not from behind or in front, where it would be more likely to cause brightness and glare problems. (See Figure 10.1.)
- Make sure that the desk is located where adequate access to power, phone, and network connections exists, and where you and others can still move unhindered throughout the work area.

Once the desk is in its final position with all power, phone, and network connections accessible, you can adjust the chair.

Step 2: Adjusting the Chair
Adjusting the chair is the physical action of optimizing seat, backrest, and arm positions. Sitting

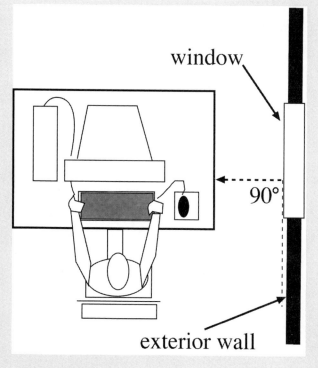

Figure 10.1
Desk placement relative to external window.

for long periods can be particularly hard on your back, legs, and feet and can impede blood circulation. So you should optimize your chair's adjustments no matter what your size and shape. This step involves setting the seat height for proper forearm position, taking further actions to ensure the feet are resting firmly on the floor, and adjusting the backrest angle to provide proper back support.

Equipment: A chair with height and back support adjustments is a must for properly setting up a computer workstation. Most adjustable chairs do not require tools and can be adjusted with hand-operated knobs and levers.

- Sit in the chair in front of the desk in your normal operating position. Make sure you and your chair fit easily under the desk. If you do not, make an initial height adjustment of the chair. Then check your arm position with your hands on the desk at the normal keyboarding position. Your forearms should be parallel to the floor with an elbow bend of 90°. Change the height adjustment so that your forearms are in the correct position. If you and your chair can no longer fit under the desk, you need a different desk or chair. (See Figure 10.2.)

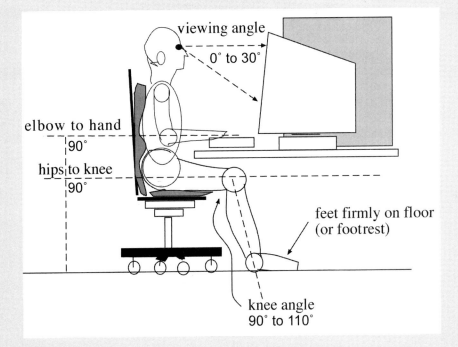

Figure 10.2
Chair adjustment.

- Ensure your feet rest firmly on the floor when your knees are at a 90° to 110° angle. If you need a footrest, get one; but do not leave your feet dangling unsupported above the floor.

* Set the backrest so that your lower back fits snugly against the back cushion, and you are either sitting upright or slightly reclining. Angle the bottom seat cushion so that your thighs are parallel to the floor with your knees and hips at about the same level.

Once the chair is adjusted, you should lay out the work area.

Step 3: Laying Out the Work Area

Laying out the work area is the physical action of optimizing the location and position of the monitor, keyboard, pointing device, and other equipment. Improperly placed monitors can cause you to bend your neck backward to read, which in turn can lead to both neck and back disorders. Also, a large percentage of health problems associated with computer use relate to the keyboard and pointing device. Extended work on computers can cause repetitive stress injuries to the shoulders, elbows, forearms, wrists, and hands. Properly positioning and angling the keyboard and pointing device can help prevent these problems.

Equipment: A monitor, keyboard with adjustable angles, and pointing device. Wrist rests also may be needed for the keyboard and pointing device.

* While you are seated at your normal work position, adjust the computer monitor so that you are the correct distance from the screen for reading with your neck in a relaxed position. Place your monitor so that the screen is 18 to 24 inches from your face. While you are sitting upright in a normal work position, if you can reach out and just touch the screen, your monitor is positioned correctly. (See Figure 10.2.)
* Adjust the monitor's height and position so that you are looking down at a slight angle when

reading the screen. For the typical monitor, the top of the screen should be at about your eye level. If you wear bifocals or progressive no-line glasses, you may need to adjust the screen level even lower so that you are not tilting your head backward when reading.

• Adjust the keyboard and mouse. Place the keyboard so that, when you are keyboarding, your wrists are straight and your forearms are still parallel to the floor. Adding negative tilt (raising the front edge of the keyboard) or using a keyboard tray under the top surface of the desk may be necessary.

• Position your mouse or other pointing device so that it is within easy reach and at the same height as the keyboard. That way, you will not have to twist your body or lean your shoulder to use it.

• Position wrist rests in front of both the keyboard and pointing device, but use them only when you are resting your hands and not actually keying or moving the mouse.

Once you have laid out your work area, you should adjust the lighting.

Step 4: Adjusting the Lighting
Adjusting the lighting is the physical action of optimizing the location, angle, and intensity of both exterior and interior illumination sources. Some of the most significant problems with computer workstations are visual fatigue, blurred vision, and headaches caused by glare and improper illumination.

External illumination sources tend to vary significantly in intensity and angle. In step 1, you already placed your desk at a 90° angle to any external window to substantially reduce glare problems from external light. However, you may still need to install blinds, shades, or curtains to block out external light sources.

Internal light sources in offices are usually too bright for computer workstation use, with overhead lighting being the single biggest problem. One solution is to adjust the monitor's angle slightly to reduce the glare. Another is to reduce the intensity of the overhead lights, by removing some of the bulbs, installing dimmer circuits where appropriate, or simply switching off banks of lights altogether. The contrast of the display can be increased, and glare screens such as neutral density and polarizing filters can be added in front of the monitor. Neutral density filters reduce the intensity of the glare relative to the intensity of the displayed information; polarizing filters block the polarity of the reflected glare while permitting the displayed information to pass. Monitor shields also can be installed around the screen to block the source of the glare.

Equipment: It varies significantly with the amount of glare and the solutions required.

- While you are seated at your normal work position, look at your computer monitor. If you see reflections, you have glare. First, try to remove it by slightly tilting the monitor. If that does not work, identify the source of the glare. Doing so is easy because you can see it reflecting in your monitor's screen.
- If an external window is the problem, you can reduce the glare with shades, blinds, or curtains.
- If internal lights are the problem, move the lights or reduce their intensity. In some cases, you can just turn them off. Use supplemental desk lighting to compensate for the loss of overhead lighting. If that does not work, install glare screens and monitor shields.

When you are able to sit in your normal work position and not see annoying reflections in your monitor's screen, the workstation's setup is complete.

Conclusion

Setting up a computer workstation is an ergonomics-based process designed to enhance the user's health, comfort, and productivity. Generally speaking, setting up a computer workstation involves optimizing the desk's location, adjusting the chair, arranging the work area, and adjusting the lighting.

For more detailed information on setting up your computer, see "Computer Workstation Ergonomics," Center for Disease Control, Internet: http://www.cdc.gov/od/ohs/ergonomics/compergo .htm.

Instructions for an Expert

The following example is geared to an expert audience and involves critical modifications to a high-power transmitter circuit. Notice the language used and the assumptions being made about the audience's skill level. The theoretical background information that the audience should know has been omitted, and specific details are provided when either unique information is required or health and safety are at risk.

Neutralizing the 16XL1000000 in Class C High-Gain Applications

Introduction

Neutralization is the process of introducing negative feedback into high-gain, power amplifier stages to prevent unwanted oscillation. The 16XL1000000 Megatube is a high-power tetrode capable of producing an input of more than a megawatt of RF power. It is used as the class C RF power amplifier in the Acme 2000XL Ultra High Power Shortwave Transmitter currently employed in military operations. In this high-gain class C application, user experience indicates that the 16XL1000000 may require neutralization. Although the tube's screen grid reduces plate-to-grid capacitance to a fraction of a picofarad, the high power gain is such that oscillation may still occur. At the megawatt level, such oscillation could lead to a catastrophic equipment failure. To prevent such oscillation, the power amplifier should be neutralized.

Neutralization requires modification of the original circuit to provide enough negative feedback to prevent oscillation. This process involves replacing the original input inductor L2 with a center-tapped input inductor L2A to provide the necessary phase-reversed access to the input signal, and then adding variable capacitor C to couple the output signal at the correct level to the input signal. The neutralization process consists of the following steps: (1) replacing the input inductor, (2) installing the high-voltage plate tap, (3) installing the neutralizing variable capacitor, and (4) adjusting the capacitor.

DANGER: THIS AMPLIFIER OPERATES AT 22 KILOVOLTS AT 60 AMPERES. DEATH CAN OCCUR ON CONTACT. Ensure the system is completely powered down, all filter capacitors are discharged, and that you follow all high-voltage precautions associated with this system. Refer to Figure 10.3 for the power amplifier's schematic with the neutralization components installed.

Step 1: Replacing the input inductor
Replacing the input inductor is the mechanical process of substituting the center-tapped L2A for the original untapped L2. One end of the center-tapped L2A provides a required phase-reversed signal relative to the other end. This out-of-

Figure 10.3 - Amplifier circuit with neutralization components installed.

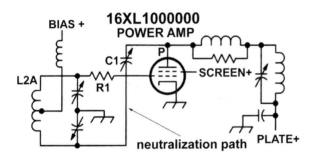

phase signal can then be used to provide negative feedback in the neutralization process.

To accomplish this step, you will need a 6-mm socket wrench for the mounting flange bolt and a Torx 12 driver for the lead connector screws. CAUTION: RF INPUT DRIVE OF 3000 WATTS IS PRESENT ON THIS INDUCTOR DURING NORMAL OPERATION. Consequently, ensure the replacement inductor is mounted in exactly the same position as the original inductor, using the existing mounting flange hardware.

DANGER: THIS AMPLIFIER OPERATES AT 22 KILOVOLTS AT 60 AMPERES. DEATH CAN OCCUR ON CONTACT. ENSURE PLATE POWER IS <u>OFF</u> AND THAT ALL FILTER CAPACITORS HAVE BEEN DISCHARGED BEFORE ACCOMPLISHING THIS STEP.

• Disconnect the leads to the existing tuning capacitors, unbolt the mounting flange, and remove L2. Next, install the replacement inductor L2A *in exactly the same position* by mounting the capacitor on the original flange, tightening the flange, and reconnecting and tightening the leads.

The replacement inductor L2A should fit precisely where the original inductor L2 was mounted, using the existing mounting hardware, and should be mounted in the same location and position as the original inductor L2.

Next, install the high-voltage plate tap.

Step 2: Installing the high-voltage plate tap
Installing the high-voltage plate tap is the mechanical process of mounting a threaded connector onto the plate electrode at P. The tap clamps onto an exposed area of the plate electrode and provides the necessary output signal for the neutralization process. CAUTION: OVERTIGHTENING THE CLAMP CAN CRACK THE CERAMIC SEAL AND DESTROY THE TUBE.

DANGER: THIS AMPLIFIER OPERATES AT 22 KILOVOLTS AT 60 AMPERES. DEATH CAN OCCUR ON CONTACT. ENSURE PLATE POWER IS OFF AND THAT ALL FILTER CAPACITORS HAVE BEEN DISCHARGED BEFORE ACCOMPLISHING THIS STEP.

• Clamp the tap onto the plate electrode P.

Ensure the clamp is snug and tight (about 30 foot-pounds) before you install the neutralization capacitor.

Step 3: Installing the neutralization capacitor
Installing the neutralization capacitor is the mechanical process of physically mounting the device and providing electrical connectivity to the circuit environment. In this step you will locate the capacitor in the unused area under the main chassis near the anode electrode of the tube. Mount the capacitor, using ceramic standoffs at least 6 centimeters long. CAUTION: FULL PLATE VOLTAGE AND CURRENT ARE AVAILABLE ON ONE SIDE OF THE CAPACITOR; MOUNT THE CAPACITOR IN A WAY THAT PRECLUDES HIGH-VOLTAGE FAULTS BY ENSURING THAT A CLEAR SPACE OF AT LEAST 20 CENTIMETERS EXISTS AROUND THE CAPACITOR AND ITS TERMINAL LEADS.

DANGER: THIS AMPLIFIER OPERATES AT 22 KILOVOLTS AT 60 AMPERES. DEATH CAN OCCUR ON CONTACT. ENSURE PLATE POWER IS OFF AND THAT ALL FILTER CAPACITORS HAVE BEEN DISCHARGED BEFORE ACCOMPLISHING THIS STEP.

• Mount the neutralization capacitor in the circuit environment, connecting one lead to L2A and the other lead to P. *Before* you adjust the capacitor, be sure to check that the capacitor is mounted and its terminal leads are routed to preclude high-voltage arcing.

Step 4: Adjusting the neutralization capacitor

Adjusting the neutralization capacitor is a calibration process in which the amount of negative feedback is set to the proper level. In this step, you will follow standard neutralization procedures to adjust the capacitor to cancel any coupled energy. The use of a low-level RF generator connected to the output with a sensitive RF detector connected to the input is recommended for this process.

DANGER: THIS AMPLIFIER OPERATES AT 22 KILOVOLTS AT 60 AMPERES. DEATH CAN OCCUR ON CONTACT. ENSURE PLATE POWER IS <u>OFF</u> AND THAT ALL FILTER CAPACITORS HAVE BEEN DISCHARGED BEFORE ACCOMPLISHING THIS STEP.

- Connect the low-level RF generator to the output and a sensitive RF detector to the input, both operating at the operational resonant frequency of the power amplifier. Maximize amplifier tuning and loading at this operating frequency, and then adjust C1 for minimum response on the RF detector.

At this point, the amplifier is neutralized.

More on Manuals

As mentioned earlier, manuals follow the same basic pattern as instructions for telling someone how to do something. However, manuals have a much larger scope and are more comprehensive.

Manuals provide instructions for many complicated tasks involving complex equipment. In some cases where systems are large and extremely complex (such as an aircraft or automobile), many separate manuals are required and may, in fact, fill a small library. Smaller systems (such as a lawnmower) frequently require only a single, self-contained manual. Increasingly, manuals are provided in electronic format, most often in the form of hypertext (.htm, .html), Adobe Portable Document File (.pdf), or word processing files such as Microsoft Word (.doc). (See Chapter 16 for a more complete discussion of electronic publishing.)

Here are the most common types of manuals and the kinds of tasks with which they are used:

- *Assembly manuals:* construction, alignment, calibration, testing, or adjusting a mechanism.
- *Owner or user manuals:* use of a mechanism, routine maintenance, and basic operation.
- *Operator manuals:* use of a mechanism and very minor maintenance.
- *Service manuals:* routine maintenance of a mechanism, including troubleshooting, testing, repairing, or replacing defective parts.
- *Technical manuals:* parts specifications, operation, calibration, alignment, diagnosis, and assembly.

Instructions Checklist

- Do I understand the process and the skill level of the intended audience?
- Have I defined the overall process and described its purpose?
- Have I explained any needed theories or principles?
- Have I listed the steps?
- For each step:
  - Have I defined the step?
  - Before I tell the reader to actually do something...
    - Have I given an overview of what will happen in this step?
    - Have I provided needed information such as cautions, dangers, and required equipment?
  - When I tell the reader to do something...
    - Have I described exactly what should be done?
    - Have I described the result that should occur?

- Have I transitioned to the next step, if there is one?
- Have I summarized all the steps?
- Have I told the reader how to get additional information?

Note   1. Special thanks to Sharon E. Liebel, CPE, Ergonomics Engineer and Management Consultant, Humantech, Inc., Ann Arbor, Michigan, for her help with this example.

11

Research Reports

Research reports are similar to research papers that every student has done at one time or another. You remember—papers with titles such as "The Sexuality of Lady Macbeth" or "The Benefits of the Modern Mosquito." In technical writing, however, research reports are focused, objective inquiries into technical subjects.

Research reports describe the discovery, analysis, and documentation of knowledge obtained through some type of investigation. In technical writing, these reports are specifically geared to the purpose at hand, the readers who will use them, the clients who will read them, and whatever limitations have been placed on the scope of the project. Technical research reports frequently focus on new, evolving, sometimes purely hypothetical technologies, in which case they are called *state-of-the-art reports*.

One distinguishing characteristic of research reports is the extensive research and documentation required. The research may consist of library and laboratory research, interviews, questionnaires, various types of corporate technical reports, and trade journal articles. Also research report writers increasingly use the wealth of information on the Internet. Unlike laboratory reports, however, research reports do not involve doing the actual research being reported; they only present the findings of research that has already been done.

What Are Research Reports?

The organization of a research report is straightforward, as shown in Outline 11.1. However, what goes in the discussion section depends on the topic and the specific requirements for the research. If the focus is on how we got to where we are in developing a certain technology, the discussion will be primarily historical. But if the purpose is to describe new, evolving technologies, the discussion may be geared more to future implications.

Research reports are usually comprehensive documents that often extend beyond the scope of this book. However, to illustrate the kinds of things that go into a research report, this chapter will provide two abbreviated examples. As in prior chapters, these examples will employ theoretically correct, but fictitious technologies. The first example is a state-of-the-art report on the QCPU, a quantum computing chip. The second is a research report on shortwave-broadcast transmission antennas for use with the 16LX1000000 tube. Both examples follow Outline 11.1.

Outline 11.1 Research Report

Introduction	
• Purpose	Describe the reason for writing this report.
• Problem	Provide a brief overview or introduction for the topic.
• Scope	Describe the limitations of this report.
Background	
• Theory	Review any theory needed to understand the topic.
• History	Provide any necessary historical perspective.
Discussion	(Main section of the report—content can vary significantly.)
Conclusion	
• Summary	Summarize the discussion section.
• Recommendation	Provide suggestions based on the summary.

References
- Sources cited   List sources consulted and used in the report.
- Sources not cited List sources consulted but not specifically used.
Appendixes   Provide supporting material not needed to understand the report.

This first example is a state-of-the-art report on Quantum Chips Corporation's Quantum Central Processing Unit (QCPU). This powerful piece of vaporware exploits the tremendous potential of quantum computing architectures to effectively increase, by several orders of magnitude, the high-end computing power currently available. (Quantum computing is actually an area of serious scientific research among leading physicists and engineers. It is an extremely dynamic theoretical field.)

Developing a Research Report

Introduction

As in other documents, the first part of the research report is the introduction section. Start with the purpose statement to explain why you are writing the report:

Purpose

The purpose of this report is to provide the results of a state-of-the-art investigation of the Quantum Chips Quantum Central Processing Unit (QCPU), including a theoretical review of the premises of its operation.

Next, state the problem that the report addresses. In a research report, note that the problem is really more of a general background statement that expands on the topic and gives a brief context for what the report will investigate. Also note the inclusion of a source citation to support the assertion regarding processing speed. This type of documentation is required in research reports and is explained more thoroughly in Chapter 14.

Problem

Traditionally, computing power has been enhanced by increasing CPU speeds primarily through decreasing the size of conductors and solid-state devices used in chip fabrication. Decreasing size, however, has finite limitations, such as those associated with reducing the dielectric constants of the required materials. There has also been a move toward increasing the number of instructions executed for each clock cycle, especially with reduced-instruction-set computer (RISC) processors.

Quantum Chips Corporation has transcended this traditional paradigm by developing the QCPU. The QCPU exploits and manipulates the quantum nuclear spin states of atoms. The QCPU is capable of performing large numbers of advanced computational tasks simultaneously, using the superimposition of multiple values encoded into the respective spin states of individual atoms. The resulting (effective) CPU speed is equal to or better than 500 gigahertz (Josephson 2004, p. 291). This effective speed makes the QCPU an ideal chip for processing-intensive tasks such as cryptographic factoring of large numbers, DNA sequencing in genetic research, and interactive, three-dimensional holographic imaging in advanced virtual reality systems.

In any research paper, you cannot possibly research everything about your topic. Human knowledge is not that simple or easy, and there is too much of it. So you will have to limit your paper by including only certain aspects of your topic. To complete the introduction, provide a scope statement that addresses this limitation. This section tells your reader what you are including in the paper and why, and it articulates the rationale for the limitations you are imposing.

Scope

The QCPU is built using proprietary information owned and protected by Quantum Chips Corporation. Consequently, this report will be limited to the general theoretical approaches underlying the QCPU architecture; it will not investigate the actual methods used by the QCPU to implement these theoretical approaches.

Background

In the background section, discuss the theoretical and historical aspects of the topic, as appropriate. Given the purpose here, this example will focus on only the theoretical aspects of Quantum Chips' QCPU technology. The background should start with a brief discussion of quantum computing theory because this theory is not common knowledge for the audience; consequently, the theory discussion is essential to understanding the rest of the paper. Note again the inclusion of source citations throughout this discussion.

Theory

Quantum computing theory applies the properties of quantum physics to exploit subatomic phenomena of common elements to perform extremely complex computational tasks. When properly exploited, these phenomena provide a truly unprecedented ability for massively parallel processing (Ardvark 2004, pp. 446–448). Several options exist to exploit quantum phenomena in this regard. One is to equate binary values to the ground and excited states of an atom. Another is to use traditional nuclear magnetic resonance (NMR) techniques to read induced spin states of atoms. A third is to polarize photons in an optical chamber. Quantum Chips Corporation has applied the second option in the QCPU, using NMR techniques to read specifically induced spin states in carbon, hydrogen, and other atoms (Josephson 2004, p. 301).

To manipulate carbon and hydrogen atoms, radio frequency (RF) energy is applied to each atom at its specific resonant frequency. This RF energy is applied to the atom while it is in a fixed magnetic field. Because the atom remains in a fixed position, the position can serve as its memory address. The nucleons of these atoms spin predictably while in this magnetic field. If an atom lines up with the direction of the magnetic field, it is considered to be in a "spin up" orientation. If it lines up in a direction opposite to the magnetic field, it is considered to be in a "spin down" orientation. Different spin states have different energy signatures for different atoms at different magnetic field magnitudes. These differences can be read by NMR sensors.

Discussion

As an example of the kinds of discussion material this type of report might contain, some information is also provided on the genesis of the QCPU. This kind of discussion would be useful for topics that deal with radically new technologies that vary significantly from traditional methods. Quantum computing is clearly such a topic.

Genesis of the QCPU

In 1998, George Yamaslute demonstrated that different RF signals cause predictably different spins for certain atoms. He also showed that the spin of these atoms can be altered by the application of different RF signals. These altered spin states then can be used symbolically to represent different values. The spin states of these atoms effectively store the values encoded by these RF signals. These values are then read by traditional NMR techniques, thereby creating a machine memory capability (Yamaslute 1998, pp. 200–210).

Besides memory capabilities, manipulation of the spin states of atoms can be used to perform various logic operations. Early experiments with molecules

containing carbon and hydrogen atoms demonstrated such theoretical feasibility. The carbon and
hydrogen atoms can be manipulated independently
by varying RF signals applied at different resonant
frequencies while these atoms are held in a fixed
magnetic field. By making both atoms spin up, or
spin down, or alternately spin up and spin down, the
atoms can constitute a 2-bit truth table. Adding
more types of atoms and using intermediate spin
states substantially increase the range of logical operations. Such manipulation provides the
quantum logic capabilities of the QCPU (Yamaslute
1998, p. 240).

Finally, the discussion should include a brief
description of the QCPU device because the chip
is the primary focus of this report.

The QCPU

The Quantum Chips QCPU is proprietary technology used today only in highly classified government
projects. Although unclassified information is limited, some scientists believe that the QCPU is now
providing the computational power for the genetic
manipulation of bacteria. The specific goal is to create virulent strains of Naomi-B bacteria that are
capable of eating through the armored titanium and
steel hulls of enemy submarines at depths exceeding
10,000 feet (Mierson 2003, p. 50).

The QCPU assembly consists of the quantum
molecular matrix (QMM), the magnetic field coil
(MFC), the nuclear magnetic resonance (NMR) sensor, and the RF assembly (RFA). The MFC provides
the fixed magnetic field engulfing the QMM. The
QMM provides atomic structures that have unique
spin up and spin down characteristics. The RFA provides the RF energy needed to change the spin characteristics of each atom to reflect specific values
(Bearkins 2000, p. 91). The energy is radiated by a
phase directional array, but the exact design or
method of control is proprietary. The NMR sensor is
the means of sensing or reading these altered spin
states.

Conclusion

The conclusion section of a research report normally summarizes the report and may provide a recommendation. Any recommendation must be supported and justified by information in the discussion section. Given the nature of this sample report's theoretical discussion, a specific recommendation is not supported or justified, and one is not provided. In this example, the conclusion will summarize only the information provided on the QCPU.

Summary

The Quantum Chips Quantum Central Processing Unit (QCPU) has made quantum computing a reality. By controlling and reading the spin states of selected atoms and using applied RF and NMR techniques, the QCPU effectively uses quantum phenomena to store data and conduct logical operations on those data. Quantum computing, by exploiting the possibilities of data manipulation and storage at the atomic level, provides the power to accomplish parallel processing at far greater amplitudes than traditional approaches.

The actual design and implementation of the QCPU is not only proprietary but also highly classified for national security purposes. Little information is available about the specifics of the QCPU's design implementation.

References and Appendix

Finally, include a list of the references used in the report. Always list all references that were actually cited in the report. As a courtesy to your reader, you can also list sources that you consulted but did not specifically use because these sources may have influenced your thinking or may provide additional information for further exploration of the topic. In this example, five sources were actually cited in the paper, and one source was consulted but not used.

Consequently, in the following section of this chapter, where the complete report has been assembled, you will note that two categories of references are listed: *sources consulted and used* and *sources consulted and not used*. For a more detailed discussion of documentation, see Chapter 14. By the way, although not provided here, this example could also include an appendix containing additional, useful information that is not necessary for understanding the report. For example, this material might include the manufacturer's press releases on the QCPU, more information on the use of NMR in materials research, or detailed discussions of atomic spin states.

Here is the complete state-of-the-art research report, including visuals, references, and appendix.

Putting It All Together

State-of-the-Art Report on Quantum Chips' Quantum Central Processing Unit

Introduction
Purpose
The purpose of this report is to provide a state-of-the-art investigation of Quantum Chips' Quantum Central Processing Unit (QCPU), including a theoretical review of the premises of its operation.

Problem
Traditionally, computing power has been enhanced by increasing CPU speeds, primarily through decreasing the size of conductors and solid-state devices used in chip fabrication. Decreasing size, however, has finite limitations, such as those associated with reducing the dielectric constants of the required materials. There has also been a move

toward increasing the number of instructions executed for each clock cycle, especially with reduced-instruction-set computer (RISC) processors.

Quantum Chips Corporation has transcended this traditional paradigm by developing the QCPU. The QCPU exploits and manipulates the quantum nuclear spin states of atoms. The QCPU is capable of performing large numbers of advanced computational tasks simultaneously, using the superimposition of multiple values encoded into the respective spin states of individual atoms. The resulting (effective) CPU speed is equal to or better than 500 gigahertz (Josephson 2004, p. 291). This effective speed makes the QCPU an ideal chip for processing-intensive tasks such as cryptographic factoring of large numbers, DNA sequencing in genetic research, and interactive, three-dimensional holographic imaging in advanced virtual reality systems.

Scope

The QCPU is built using proprietary information owned and protected by Quantum Chips Corporation. Consequently, this report will be limited to the general theoretical approaches underlying the QCPU architecture; it will not investigate the actual methods used by the QCPU to implement these theoretical approaches.

Background

Theory

Quantum computing theory applies the knowledge of quantum physics to exploit subatomic phenomena of common elements to perform extremely complex computational tasks. When properly exploited, these phenomena provide a truly unprecedented ability for massively parallel processing (Ardvark 2004, pp. 446–448). Several options exist to exploit quantum phenomena in this regard. One is to equate binary values to the ground and excited states of an atom. Another is to use traditional nuclear magnetic resonance (NMR) techniques to read induced spin states of atoms. A third is to polarize photons in an optical chamber. Quantum Chips Corporation has applied the second option in the QCPU, using NMR techniques to read specifically induced spin states in carbon, hydrogen, and other atoms (Josephson 2004, p. 301).

To manipulate carbon and hydrogen atoms, radio-frequency (RF) energy is applied to each atom at its specific resonant frequency. This RF energy is applied to the atom while it is in a fixed magnetic field. Because the atom remains in a fixed position, the position can serve as its memory address. The nucleons of these atoms spin predictably while in this magnetic field. If an atom lines up with the direction of the magnetic field, it is considered to be in a "spin up" orientation. If it lines up in a direction opposite to the magnetic field, it is considered to be in a "spin down" orientation. Different spin states have different energy signatures for different atoms at different magnetic field magnitudes. These differences can be read by NMR sensors.

Discussion
Genesis of the QCPU

In 1998, George Yamaslute demonstrated that different RF signals cause predictably different spins for certain atoms. He also showed that the spin of these atoms can be altered by the application of different RF signals. These altered spin states then can be used symbolically to represent different values. The spin states of these atoms effectively store the values encoded by these RF signals. These values are then read by traditional

NMR techniques, thereby creating a machine memory capability (Yamaslute 1998, pp. 200–210).

Besides memory capabilities, manipulation of the spin states of atoms can be used to perform various logic operations. Early experiments with molecules containing carbon and hydrogen atoms demonstrated such theoretical feasibility. The carbon and hydrogen atoms can be manipulated independently by varying RF signals applied at different resonant frequencies while these atoms are held in a fixed magnetic field. By making both atoms spin up, or spin down, or alternately spin up and spin down, the atoms can constitute a 2-bit truth table (Table 11.1). Adding more types of atoms and using intermediate spin states substantially increase the range of logical operations.

Table 11.1 Spin state truth table

RF signal applied	Carbon	Hydrogen
Frequency 1	U	U
Frequency 2	U	D
Frequency 3	D	U
Frequency 4	D	D

Such manipulation provides the quantum logic capabilities of the QCPU (Yamaslute 1998, p. 240).

The QCPU

The Quantum Chips QCPU is proprietary technology used today only in highly classified government projects. Although unclassified information is limited, some scientists believe that the QCPU is now providing the computational power for the genetic manipulation of bacteria. The specific goal is to create virulent strains of Naomi-B bacteria that are capable of eating through the armored titanium and steel hulls of enemy submarines at depths exceeding 10,000 feet (Mierson 2003, p. 50).

The QCPU assembly consists of the quantum molecular matrix (QMM), the magnetic field coil (MFC), the nuclear magnetic resonance (NMR) sensor, and the RF assembly (RFA). (See Figure 11.1.) The MFC provides the fixed magnetic field engulfing the QMM. The QMM provides atomic structures that have unique spin up and spin down characteristics. The RFA provides the RF

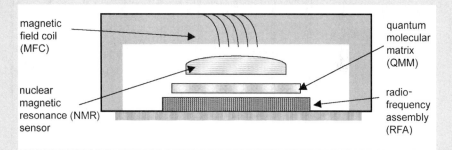

Figure 11.1
QCPU assembly. [*Source: George S. Yamaslute, "Magnets and Bits,"* Journal of Applied Magnetic Resonance *4:9 (September 2002), p. 260.*]

energy needed to change the spin characteristics of each atom to reflect specific values (Bearkins 2000, p. 91). The energy is radiated by a phase directional array, but the exact design or method of control is proprietary. The NMR sensor is the means of reading these altered spin states.

Conclusion
Summary
Quantum Chips' Quantum Central Processing Unit (QCPU) has made quantum computing a reality. By controlling and reading the spin states of selected atoms, using applied RF and NMR techniques, the QCPU effectively uses quantum phenomena to store data and conduct logical operations on those data. Quantum computing, by

exploiting the possibilities of data manipulation and storage at the atomic level, provides the power to accomplish parallel processing at far greater amplitudes than traditional approaches.

The actual design and implementation of the QCPU is not only proprietary but also highly classified for national security purposes. Little information is available about the specifics of the QCPU's design implementation.

References
Sources Consulted and Used

Ardvark, William J.: "NMR and Beyond," *Quantum Computer Quarterly* 9:6 (August 2004), pp. 460–495.

Bearkins, Felinus S.: "Cracking the Secrets of the Q-CHIP," Internet: http://www.crackerbox.org, March 31, 2000.

Josephson, Albert E.: "Technical Report 1999-35." Quantum Chips International, November 2004.

Mierson, Wilhelm F.: "The Dreaded Threat of Naomi-B," *Journal of Truth in Science* 12:2 (February 2003), pp. 460–462.

Yamaslute, George S.: "Magnets and Bits," *Journal of Applied Magnetic Resonance* 4:9 (September 1998), pp. 150–265.

Sources Consulted and Not Used

Baker, Joe-Bob.: *A View of That Which We Cannot See*. New York: Acme Press, 1994.

The following example again involves that rock group of all rock groups, the Village Thumpers. The high-power audio system for what became known as "Thumper Stadium" (Chapter 9) was so successful that the group purchased a large block of shares in *Boss*RF, the manufacturer of the 16XL1000000 megawatt transmitting tube used in the stadium's audio system. The group now wants to share its music with the world via high-power, shortwave broadcast through the Thumper Shortwave Network (TSN). The "Thumpers" recently purchased a Big Signal Corporation X1E6 shortwave transmitter that uses the 16XL1000000 to produce 1.2 megawatts of continuous, high-level, amplitude-modulated (A3) radio-frequency (RF) energy in the high-frequency (HF) spectrum. They also arranged for the proper licensing through improper political influence, and they have acquired 200 acres of land near Dayton, Ohio, as the transmitter site. Their technical consultants now need to select the type of antenna system that will be used and have requested a short research report that presents antenna considerations for their 16XL1000000-based transmitter.

Antenna Considerations for Shortwave Transmitters Using the 16XL1000000

Introduction

PURPOSE

The purpose of this report is to provide information regarding antenna systems typically used in high-power shortwave broadcasting that would be suitable for use with the 16XL1000000 high-power transmitting tube.

PROBLEM

Shortwave transmitters employing the 16XL1000000 produce more than 1 megawatt of power in the 49-, 31-, 25-, 19-, and 16-meter HF bands. Antennas are designed not only to handle the high RF voltages and currents associated with these power levels, but also to exhibit the proper impedance and provide optimal propagation characteristics for the coverage desired.

SCOPE

This study is limited to antenna considerations for long-range, evening propagation in the 31- and 49-meter bands (9.5 to 9.9 megahertz and 5.95 to 6.2 megahertz, respectively) using a final amplifier input power exceeding 1 megawatt.

Background

THEORY

Shortwave broadcast antennas for the 31- and 49-meter bands are designed to take advantage of propagation characteristics through the F layer of the ionosphere. The F layer forms from a combination of F1 and F2, the two upper layers of the ionosphere. The F1 and F2 layers, which range between 125 and 250 miles above the earth, combine during the evening hours, forming a higher-density layer that reflects shortwave broadcast energy back to earth. Often F layer propagation is referred to as *skywave* or *skip* (Noikins 2004, p. 296).

Different frequency bands exhibit varying skip propagation characteristics depending on time of day and sunspot activity. Shortwave broadcast antenna systems must be capable of adapting to these variations to ensure required broadcast coverage. Consequently, antennas must have the capability of varying the angle of radiation in the vertical plane to hit the ionosphere at the proper angle for the reflections to reach the target geographic area. Likewise, antennas must have control over the angle in the azimuthal plane to direct the energy horizontally toward the target geographic area.

Additionally, shortwave broadcast antennas at these power levels must be almost totally efficient (I^2R losses <1 percent), operate with low reflected power (SWR < 1.3:1), and provide relatively high gain (10 to 25 decibels) in transmitted lobes (Noikins 2004, pp. 278–296). Proper impedance matching and tuning of these antenna systems are critical, as is the physical location of the array relative to surrounding structures.

HISTORY

Traditionally, several different antenna systems have been used for shortwave broadcasting. For lower-power systems (< 100 kilowatts), the two most popular are the *log-periodic antenna* (*LPA*) and *fan dipole antenna* (*FDA*). LPAs are produced in many configurations and can be used as either omnidirectional or rotating directional arrays. Typically, these antennas provide a beam width of 75° and a gain of 10 decibels. LPAs can be used at power levels up to 100 kilowatts. FDAs are used only for omnidirectional service at power levels less than 50 kilowatts (Fischlett 2003, pp. 465–471).

For high-power service, and especially for applications involving the extreme power levels of the 16XL1000000, the dipole curtain antenna (DCA) is most frequently used (Figure 11.2). These antennas employ rows and columns of dipoles arrayed in front of a reflecting screen. Although they have limited beam widths

($< 30°$), extensive field switching permits control over the angle of radiation in both the vertical and azimuthal planes.

Figure 11.2 - Dipole curtain antenna.

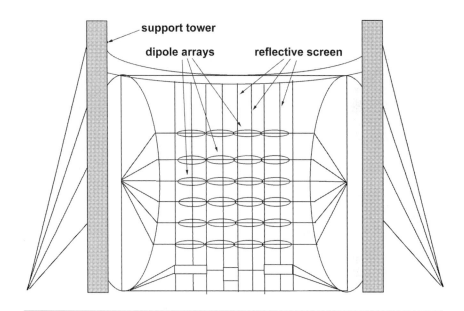

Discussion

COVERAGE CONSIDERATIONS

The DCA is the overwhelming favorite for international broadcasters and is the appropriate antenna system for operation at 16XL1000000 power levels (Peudawg 2001, pp. 21–28). To ensure effective broadcast of Thumper music to most populated areas, preliminary analysis indicates that the Dayton, Ohio, location would require a DCA employing four columns and six rows of dipole arrays suspended on both sides of a single reflecting screen. The screen would be suspended in a north-south orientation (Zigwee 2004). One set of dipole arrays would service points to the east, while a second set on the other side of the screen would service points to the west. This configuration would provide a beam width of about 28° (Peudawg 2001, p. 171). Through column field switching, the beam's directional lobes could be slewed by 30°, providing 88° of effective coverage in each direction. The rows also can be switched to control the beam's angle of radiation in the vertical plane. Additionally, natural offset lobes at 90° to the primary lobes will provide some coverage to the north and south (Figure 11.3).

Figure 11.3 - Coverage map.

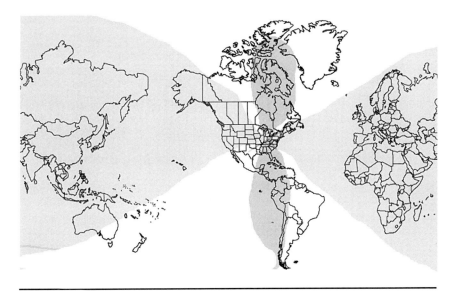

IMPEDANCE CONSIDERATIONS

The DCA exhibits a typical impedance of 300 ohms. The antenna should be fed with a 300-ohm balanced, open transmission line suitable for the RF voltages and currents resulting from 1.2 megawatts of power. To match the transmitter's output impedance of 50 ohms, a tapered-line balun will be required. Additionally, the transmitter will require a well-shielded antenna tuning unit (ATU) consisting of a tuned, high-voltage, helix coil (rated at 500 kilovolts rms) with a grading corona ring and a protective gap to counter lightning-induced transients. The ATU will employ sets of variometers to provide continuous, fine adjustment of inductance, and fixed gas capacitors to provide the required static capacitance.

ENVIRONMENTAL CONSIDERATIONS

Very high electromagnetic fields have been correlated with a variety of physiological problems, such as difficulty in sleeping, reproductive disorders, and interference with audio and other electronic devices. For example, in some cases, poorly designed audio equipment, decaying dental amalgams, or copper plumbing with deteriorated joints has effectively demodulated high-power shortwave transmissions, often creating discomfort or irritation for the affected parties. However, given the acreage available for this system, the relative remoteness of the antenna site, and the fact that high-intensity electromagnetic fields at these frequencies have not been shown to *cause* injury or death, environmental impact is not considered a problem for the purposes of this study.

Conclusion

SUMMARY

Construction and operation of the Thumper Shortwave Network (TSN) transmitter site near Dayton, Ohio, appears to be viable in terms of antenna considerations. The DCA would be the appropriate choice, with an array of four columns and six rows of dipole cages positioned on each side of a reflective screen. The antenna would provide the gain, azimuthal slewing, and vertical elevation (takeoff angle) control to cover most of the major population centers of the earth, using the appropriate shortwave bands. Use of an ATU, balun, and balanced transmission line would effectively match the DCA to the 16XL1000000 power amplifier. Environmental considerations are not deemed significant for the purposes of this study, but additional research may be warranted before a final decision on this project is reached.

RECOMMENDATION

This report recommends that the project move forward with a full and complete analysis of the Dayton, Ohio, site to ensure technical, economic, and environmental feasibility.

References

SOURCES CITED

Fischlett, Robert F.: *Antenna Types and Advantages.* London: RF Press, 2003.
Noikins, Nikonius: "Antenna Matching Techniques," *Journal of Applied Antenna Philosophy,* 9 (September 2004), pp. 270–315.
Peudawg, Alicia M.: *International Broadcasting Made Easy.* New York: Airwaves Press, 2001.
Zigwee, Alan B.: "Thumper Report on Shortwave Antennas," RFSPL-04-1243. Williston, N.Y.: RF Signals and Propagation Laboratories, Inc., Williston, N.Y., January 2004.

Research Report Checklist	• Have I clearly stated the purpose of this report? • Have I introduced the topic with a brief overview of the problem or background? • Have I discussed how I limited the report and my rationale for doing so? • Have I provided adequate background for my reader to understand the report? • Have I provided substantive, well-documented information in the report? • Have I included necessary visuals and data? • Have I summarized my research in the conclusion? • Have I made a recommendation; and if so, is it supported by the discussion? • Have I cited sources where necessary in the text? • Have I listed the sources I cited? • Have I included in an appendix any relevant material not necessary for understanding the paper?

12

Abstracts and Summaries

When you have finished putting together your technical report, you still may not be finished writing. You may need to add a special summary of your report, called either an *abstract* or an *executive summary*. Abstracts and executive summaries do the same thing: They summarize what is in your report. Abstracts are usually shorter than executive summaries and generally come in two forms: descriptive and informative.

What Are Descriptive Abstracts?

Descriptive abstracts (also called *limited abstracts*) summarize the structure of a report, but not its substance. In other words, descriptive abstracts basically present the table of contents in paragraph form. They refer to the title and the author and may briefly sketch out the purpose, problem, and scope of the document. They also describe the major topics covered by the report. A typical descriptive abstract contains around 50 words.

Writing Descriptive Abstracts

Writing descriptive abstracts is simple because you do not have to get into the substance of the report. Here is a descriptive abstract for the sample research report on Quantum Chips Corporation's Quantum Central Processing Unit (QCPU) in Chapter 11:

Descriptive Abstract

**Quantum Chips Corporation's Quantum
Central Processing Unit (QCPU)**

This research report provides a state-of-the-art investigation and theoretical review of Quantum Chips' QCPU device. The report begins with a description of its purpose, problem, and scope. Next it provides theoretical background on the device's operation. The main discussion deals with the genesis of the QCPU and provides nonproprietary information on the main parts and basic operation of the device.

Notice how this abstract provides no substantive details about the report except for general topic areas. For example, you know from the abstract that the report discusses the purpose, problem, and scope—but you do not know what the purpose, problem, and scope are. You also know that the report addresses the theoretical background regarding the chip's operation, but the abstract gives no information on this theory. In some descriptive abstracts, a brief statement providing additional details about the purpose and scope may be included.

What Are Informative Abstracts? *Informative abstracts* (also called *complete abstracts*) actually summarize the substance of your report, not just the structure. They provide a condensed discussion of the important points. In other words, informative abstracts tell the reader not only the major topics of the report, but also, in a nutshell, what you said about those topics. Informative abstracts may be designed as stand-alone documents.

If you intend for the abstract to be part of your report, you need not include the title or author of the report in the abstract. However, if you intend for the abstract to replace the report for some readers, you obviously should include the title (and author, if appropriate) at the beginning of the abstract. In either case, informative abstracts normally contain 100 to 200 words, or less than a single page of double-spaced text.

Writing Informative Abstracts

Writing informative abstracts is a little more involved than producing descriptive abstracts because you need to summarize the substance of the report. That means you need to understand the substance of the report. Never try to abstract a report that you do not fully understand. Also, decide early how you are going to organize the abstract. Normally informative abstracts are developed around the main topics of the report. For each of these topics, choose what information to include.

Here is an informative abstract of Chapter 11's report on Quantum Chips' QCPU.

Informative Abstract

Quantum Chips' Quantum Central Processing Unit (QCPU)

This research report provides a state-of-the-art investigation and theoretical review of Quantum Chips' Quantum Central Processing Unit. Quantum Chips Corporation has transcended traditional CPU designs by using quantum nuclear spin states of atoms to store and manipulate large amounts of binary data in parallel. CPU speeds equal to or better than 500 gigahertz have been realized with this chip. The device has been developed using quantum computing theory to exploit subatomic phenomena of common elements to perform extremely complex computational tasks, resulting in massively parallel processing.

The QCPU uses nuclear magnetic resonance (NMR) techniques to read specifically induced spin states in carbon, hydrogen, and other atoms. Data are stored and manipulated by using radio-frequency (RF) energy to alter the spin states of these atoms while they are trapped in a fixed magnetic field. Different spin states have distinct energy signatures for specific atoms at varying magnetic field magnitudes. These differences are read by the NMR sensor, thereby providing an accessible memory for the data stored as spin states. Besides this memory function, manipulation of the spin states of atoms also can be used to perform various logic operations.

Building on previous quantum CPU successes with carbon and hydrogen atoms, the QCPU uses proprietary technology and is built around the following: a Quantum Molecular Matrix, which provides the atomic material; an NMR sensor, which reads the spin states of the atoms; an RF assembly, which provides phased-array RF illumination of individual atoms; and a magnetic field coil, which establishes the required fixed magnetic field.

Notice the substantive details included in the informative abstract. Also, notice how some materials, such as those related to George Yamaslute and the genesis of the device, have been omitted from the abstract. The fact that Yamaslute demonstrated the feasibility of quantum computing in 1998 was not deemed important enough to be included in the abstract's summary of the report's key points. Writing these kinds of abstracts often requires some tough decisions about what stays and what goes.

What Are Executive Summaries?

An *executive summary* is normally used with large technical reports, such as formal proposals, and other fully developed business or technical documents. Executive summaries are extended, stand-alone abstracts that have both informative and descriptive characteristics. They contain both the substance and the structure of the report. In fact, an executive summary often substitutes for the full report. It is analogous to the class review notes one picks up in the bookstore to avoid buying and reading the course text. Of course, teachers always say that these notes are no substitute for the real book, and they are not—but, as we all know, they can and do work to some extent depending on the class and the teacher.

In a similar vein, executive summaries are designed to provide key management and staff with enough information about what is in a report that these executives can make informed decisions without reading the entire document. They can always go back and read the document, if warranted—or, more likely, have experts on staff read and analyze the complete report.

Executive summaries can be large documents. Major, formal, multivolume proposals often have executive summaries of 30 pages or more. In fact, a common rule of thumb is that the length of an executive summary is about 10 percent of the length of the report it summarizes. Also, because these summaries often take the place of the report for key decision makers, executive summaries can take on critical importance and must be well written.

Writing Executive Summaries

Writing executive summaries is a challenging undertaking. Your task is to capture as much of the full report's substance as is reasonable, or feasible, in a fraction of the full report's space. To do so, you need a good strategy, especially if the report you are summarizing is complex and extensive. You cannot possibly include everything, so you have to think through exactly what to include and what to leave out of the summary.

A good approach for doing that is to go back to the original purpose of the report. Evaluate everything in the report in terms of (1) how much it contributes to achieving that purpose and (2) how extensive the treatment would have to be if you were to include it. Then include only those portions that contribute the most to the goal of the report and that can be handled effectively in a summary.

As an example of an executive summary, this chapter will summarize this entire book—and, in the process, provide you with a ready-made set of review notes for the book. What is the strategy for boiling down all this material into a few pages? Obviously, not everything in this book can be included. Some guiding principle for what to include needs to be defined before we start writing the summary. As already mentioned, a good idea is to go back to the purpose of the original document. In this case we need to analyze the material in terms of how much it contributes to achieving the goal of this book, which is to explain to engineering and science students how to prepare and edit their technical writing.

Consequently, the strategy is clear: Include only the information from each chapter that discusses how to do something. That means excluding the QuadFINKEL figure skating jump, the orbital transfer maneuver study, 16XL1000000 Megatube laboratory analysis, and all the other examples. Their function in the book is to complement the "how to" discussions. Only the how-to discussions need to be in the summary. Using this focus, we will develop the following executive summary by synopsizing each chapter, beginning with the first chapter and working through the entire book.

Executive Summary
Pocket Book of Technical Writing
for Engineers and Scientists

By Leo Finkelstein, Jr.

The primary purpose of this book is to provide the necessary basics in technical writing so that engineers and scientists can produce and edit technical reports. This 19-chapter book is organized generally into three major sections: technical writing basics, technical documents, and other useful stuff.

Technical Writing Basics

Chapter 1: Introduction

Technical writing is a fundamental skill for virtually anyone working in science and engineering. Most science and engineering activities produce technical reports either on paper or in electronic form. Technical writing is the means by which these documents are produced.

Technical writing is not creative writing. It is the low-abstraction, high-precision communication of complex technical and business concepts. It is also audience- and situation-relative. The writer must ensure that the reader precisely understands the intended meaning for the purpose at hand. The audience and purpose are almost always well defined for the writer by the technical writing situation.

Technical writing has several unique characteristics that distinguish it from other types of writing. It deals with technical information. It relies heavily on visuals. It uses numerical data to precisely describe quantity and direction. It is accurate and well documented. It uses headings and subheadings for transitions. And it is grammatically and stylistically correct.

Chapter 2: Ethical Considerations

Fundamentally, ethical technical writing means using communication skills and resources with the intention of doing good.

Traditionally, ethics in technical writing has been approached by laying out sets of rules for what makes an ethical technical writer. These rules have included such things as being accurate and honest, not substituting speculation for fact, not hiding the truth with ambiguity, acknowledging the sources of ideas, complying with copyright laws, not lying with statistics, and not injecting personal bias into reports. The problem with the traditional approach is that such rules do not get at the essence of ethics: being good and doing good. If you intend to use technical writing to be good and do good, where good is defined by your society and culture, then you are acting ethically.

Plagiarism is a major ethical issue in technical writing. It is an act of theft in which you steal another person's idea, or his or her expression of an idea, and then represent it as your own. The Internet provides a plagiarist with quick access to sources and documents with virtually no risk of detection. Some who would not have plagiarized because of fear of detection do so today by using Internet resources, resulting in more plagiarists. As the numbers of plagiarists increase, plagiarism becomes more acceptable—because "everyone is doing it." Fortunately, most students are not plagiarists, and the same Internet resources that enable plagiarism can also be used to detect it.

Image compositing is another area of ethical concern because of the ease of altering photographs, the quality and credibility of "doctored" images, and the potential for misleading readers. Modifying images is appropriate in technical writing when the intent is to clarify or complement, but it is unethical when the goal is to misrepresent or deceive.

Chapter 3: Technical Definition

Virtually any kind of technical writing includes one or more technical definitions. A technical writer must be able to define terms, whether these terms refer to mechanisms or processes. The process of defining involves placing the term into a classification and then differentiating it from other terms in that same classification.

Classifying the term is often the most difficult part of defining it. The class should be a general category in which the term fits, but it cannot be too general. Normally the classification is slightly higher in abstraction than the term itself. Differentiating involves narrowing the meaning of the term to just one possibility within the class.

Chapter 4: Description of a Mechanism

Technology involves mechanisms. Being able to describe these mechanisms precisely and accurately in a way the reader can understand is perhaps the most essential skill of writing technical reports.

Mechanism descriptions are accurate portrayals of material devices with two or more parts that function together to do something. These descriptions focus on the physical characteristics or attributes of a device and its parts. These documents are built around precise descriptions of size, shape, color, finish, texture, and material. Such descriptions also normally include figures, diagrams, or photographs that directly support the text discussion.

The introduction defines the overall mechanism, describes its function and appearance, and lists the parts to be discussed. The discussion section addresses each part by first defining it and then providing detailed descriptions of the part's function and appearance. At the end of each part's discussion is a transition to the next part. The conclusion then summarizes the mechanism's function and relists the parts.

Chapter 5: Description of a Process

Process descriptions are similar to mechanism descriptions, but they focus on the unfolding steps of a process, not physical attributes. Process descriptions can deal with either the operation of mechanisms or the steps of conceptual processes.

The introduction includes a definition of the overall mechanism or conceptual process, descriptions of its purpose and function, and the steps of the process to be discussed. The discussion section first defines each step and then provides detailed descriptions of what happens during the step. At the end of each step's discussion is a transition to the next step. The conclusion then summarizes the mechanism's or process's function and relists the steps.

Technical Documents

Chapter 6: Proposals

Proposals are among the most important documents because they obtain contracts, grants, and jobs. Proposals are specialized, technical business documents that offer persuasive solutions to problems. Unlike other technical documents, proposals need to be more than objective and clear—they need to sell the reader on some idea.

All proposals must do three things: (1) describe, identify, or refer to a problem; (2) offer a viable solution to the problem; and (3) show that the proposing person or organization can effectively implement this solution. Proposals can be *formal* or *informal*, and *solicited* or *unsolicited*.

Informal proposals include an introduction that specifies the purpose, problem, and scope. They include a discussion section that describes the proposed approach and the benefits that will result from its implementation; they may provide a statement of work that lays out the tasks to be performed. In the resources section, the proposal describes the personnel, facilities, and equipment required to implement the solution. In the costs section, it presents the fiscal and time resources needed to implement the solution. In the conclusion, the proposal summarizes the benefits and risks of adopting the proposed solution and provides a contact for more information.

Chapter 7: Progress Reports

Progress reports, which are also called *status reports* or *milestone reports,* follow up accepted proposals by documenting the status of a project. They focus on various tasks that make up the project and analyze the progress that has been made on each task.

Progress reports contain an introduction that covers the purpose, background, and scope of the project. In the status section, the report analyzes tasks that have been completed and provides the status on tasks that are remaining. The conclusion appraises the project's current status, evaluates progress made, forecasts when the project will be completed, and provides a contact for more information.

Chapter 8: Feasibility and Recommendation Reports

Feasibility reports, also called *recommendation reports,* are documents that focus on solving problems. They either determine the feasibility of solving a problem in a particular way or recommend which of several options for solving a problem is the best approach.

The introduction reviews the purpose of the report, the problem that needs to be solved, and the scope of alternatives and criteria. The discussion section is organized around criteria. For each criterion, the report explains the criterion and the data collected and interprets the relative value or effectiveness of each alternative solution based on this criterion. The conclusion summarizes the data and interpretations for all criteria and all candidate solutions. A feasibility report concludes to what extent the solution is feasible. A recommendation report concludes which solution is best and makes a recommendation based on this conclusion.

Chapter 9: Laboratory and Project Reports

Laboratory reports and *project reports* present information that relates to the controlled testing of a hypothesis, theory, or device, using test equipment (apparatus) and a specified series of steps employed to perform the test (procedure). These reports explain the design and conduct of the test, how the variables were controlled, and what the resulting data show.

Laboratory reports are usually research-oriented documents that start with a hypothesis or theory to be tested. Project reports are often task-oriented documents that start not with a hypothesis, but with requirements of a project assignment. Instead of hypothesis validation, these latter reports explain how the tasks were accomplished and assess the success of a project.

The introduction includes the purpose of the report, the hypothesis or requirement that forms the problem, and the scope or limitations of the report. The background reviews relevant theory and past research. The test and evaluation section describes the apparatus and procedure used. The findings section provides the data resulting from the test and interprets these data. The conclusion provides an inference based on these interpretations and a recommendation based on this inference.

Chapter 10: Instructions and Manuals

Instructions are process descriptions for human involvement. They not only describe the steps of the process, but also show someone how to accomplish these steps. Consequently, instructions have to describe a process accurately, and they have to show a reader how to accomplish each step safely and effectively.

Instructions follow the general format of process descriptions but include additional material. The introduction is basically the same. The discussion section is different, however. For each step of the process, it (1) defines the step; (2) gives an overview of what happens in the step; (3) provides needed information specific to the step, such as dangers, cautions, and required equipment; (4) gives specific instructions for accomplishing the step; (5) shows the result that should occur; and (6) provides a transition to the next step (if there is one). The conclusion summarizes the steps of the process and tells the reader where to find additional information.

Chapter 11: Research Reports

Research reports describe the discovery, analysis, and documentation of knowledge through some type of investigation. They frequently focus on new, evolving, sometimes purely hypothetical technologies, in which case they are called *state-of-the-art reports*. Research reports are characterized by extensive research and documentation.

The introduction reviews the purpose, problem, and scope of the report. The background section reviews the theoretical basis for understanding the topic and provides a historical perspective of the topic. The discussion presents the main body of research. The conclusion summarizes the material in the discussion and provides recommendations or suggestions based on this summary. The references section includes, at a minimum, sources cited and used. The appendix contains additional supporting material not needed to understand the report.

Chapter 12: Abstracts and Summaries

In many cases, technical documents are not complete without a separate section that summarizes the structure or substance of the document. *Abstracts* and *executive summaries* provide this function—they summarize the report.

Descriptive abstracts summarize the structure of a report, but not the substance. They refer to the title and author; briefly sketch out the purpose, problem, and scope of the document; and list the major topics of the report. Descriptive abstracts are about 50 words long.

Informative abstracts summarize the substance of a report, not just the structure. They provide a condensed discussion of the important points. Informative abstracts may be designed as stand-alone documents or as a section contained within the report. Informative abstracts are 100 to 200 words long.

Executive summaries (an example of which you are now reading) are extended, stand-alone abstracts with both informative and descriptive qualities. Executive summaries are used to condense large reports such as formal proposals. They provide key decision makers with enough information to make informed decisions regarding the original report without their actually having to read the entire document. Executive summaries vary considerably in length, but they can be quite large—often 10 percent of the length of the report being summarized.

Other Useful Stuff

Chapter 13: Grammar and Style

Grammar is a set of rules providing commonly accepted standards for assembling words so that, together, they make sense and convey meaning. Style is the choice of words and the way we apply the rules of grammar in our writing.

In technical reports, most common grammar and style problems involve the following: comma splices, fused sentences, sentence fragments, misplaced modifiers, passive voice problems, verb agreement errors, pronoun agreement errors, pronoun reference errors, case errors, spelling errors, and imprecision.

Chapter 14: Documentation

Proper documentation is an essential element of technical writing—an element that can, and often does, have serious legal, ethical, and credibility implications for those who do not document correctly.

Documentation, in its general meaning, simply refers to creating documents. This chapter focuses on the specific meaning of giving formal credit to a person, organization, or publication for an idea or information that either is not original or is not common knowledge of the field. Documentation is required to meet legal requirements of copyright law, adhere to academic standards, and establish credibility.

Many styles of documentation exist, including notational and parenthetical approaches. Most technical documents use parenthetical references in the text keyed to a list of references at the end of the report.

Chapter 15: Visuals

In communicating complex topics in precise ways, technical writers rely heavily on visuals, which are powerful communication tools that can pack a huge amount of information into a small space. *Visuals* in technical writing include figures, diagrams, drawings, illustrations, graphs, charts, schematics, maps, photographs, and tables.

Use visuals only when you have a specific reason to do so—when they can directly clarify or otherwise enhance your text discussion. Document visuals with source lines, and design your visuals for reproducibility, simplicity, and accuracy.

Chapter 16: Electronic Publishing

Electronic publishing involves the electronic distribution of printed material and the elaborate networking of information using computers. Technical writing is a big part of this information revolution.

Electronic publishing includes the simple distribution of traditional documents digitally, often in the form of word processing or document description files. The more complex form of electronic publishing uses hypertext documents encoded with standardized, cross-platform markup languages. Hypertext documents, which can have extensive linking and searching capabilities, are distributed over computer networks, particularly the Internet, as well as via dedicated storage media, particularly CD-ROMs and DVDs. Many file formats exist. The most common document formats are .pdf, .doc, .htm, .tex, .rtf, and .txt. The most common graphics formats are .eps, .tif, .jpg, .gif, .png, .bmp, and .pct.

Electronic documents are similar in content to traditional technical documents, but they often are organized into parallel, interrelated structures. This approach differs from the more linear method used in traditional publishing, where paper pages must be assembled in order. Converting traditional documents to hypertext documents involves applying hypertext language tags to the text and reorganizing the material for parallel access.

Chapter 17: Presentations and Briefings

Technical *presentations* and *briefings* should be built around substantive information and ideas that are logical, coherent, and well organized. Technical presentations and briefings should also use language appropriate for the audience and should be delivered effectively.

Technical speaking situations generally fall into one of three categories: impromptu presentations, in which you have no warning or preparation time; extemporaneous presentations, which are well prepared and rehearsed, but not scripted; and manuscript presentations, in which you read a prepared script. Extemporaneous is the preferred choice.

Technical speaking purposes also generally fall into one of three categories: informative, in which the primary goal is to give an audience facts and other information; demonstrative, in which you show the audience how to do something or how something works; and persuasive, in which you try to convince the audience to make a particular decision or take some specific action.

Technical briefings are focused oral presentations that use visual aids referred to as charts. Charts can take the form of slides, view graphs, or computer-generated graphics. A typical technical briefing includes a title chart, overview chart, discussion charts, summary chart, and concluding chart.

Chapter 18: Resumes and Interviews

Resumes are specialized proposals by which you offer your services to fill a position. Engineering- and science-related resumes include your name, address, telephone number, and maybe an e-mail address. The objective describes the specific kind of position desired. The strengths section points out your strongest skills and attributes. The education section documents your formal education, including degrees, certifications, honors, and relevant course areas. The computer skills section documents your computer literacy, including operating systems, languages, applications, and platforms. The experience section lists your job experience. The personal section includes job-related personal information that enhances your value to the prospective employer.

Cover letters are used to send resumes to a prospective employer. They should demonstrate your knowledge of the employer and job, briefly summarize your skills and experience, describe your desirable personal traits, and invite a favorable response.

Resumes and cover letters can get you an *interview,* and an interview can get you a job. When you report for an interview, bring all the information necessary to fill out a job application, along with extra copies of your resume. Also, fully research the employer before the interview, and be prepared to answer both technical and personal questions. Finally, be prepared to discuss salary requirements, request clarification for interview questions you do not understand, and look professional and act professional at all times.

Chapter 19: Team Writing

Team writing, often referred to as *group writing,* is the process whereby two or more authors work together to produce a document or set of documents to fulfill a requirement. When properly managed, the team members can be a powerful force; when not managed properly, members can waste resources and produce poor quality.

In student team writing, small groups of students are assigned, for educational purposes, a course project that requires a written report. In professional team writing, selected employees with relevant expertise and experience are assigned to a project because of the complexity of work and competitive, real-world requirements.

The process of team writing varies significantly from one job to another. The requirements of the situation and the resources available usually determine how teams are formed, leaders are designated, and tasks are assigned. Generally, however, the team writing process includes the following steps: determining requirements, taking preliminary actions, and producing the document.

Conclusion

Abstracts and *executive summaries* summarize technical reports. Abstracts can be either descriptive or informative. *Descriptive abstracts* summarize the structure of a report, but not its substance. *Informative abstracts* summarize the substance of a report, not just the structure. Descriptive abstracts are often about 50 words in length, while informative abstracts usually run about 100 to 200 words.

An *executive summary* provides a comprehensive, detailed synthesis of a large technical report. This type of summary is intended as a stand-alone document that can substitute for the report itself. An executive summary is designed to provide enough information about what is in the report to allow informed decisions without the need to read the report.

CHAPTER THIRTEEN

Grammar and Style

This chapter is not intended as a grammar lesson. Rather, the goal is to provide a practical guide to common grammar and style errors in technical reports—and specifically, to help you identify and fix these problems when you are proofing and editing.

Grammar: What Is It and Why Is It a Big Deal?

Grammar is nothing more than a large set of rules—commonly accepted standards for assembling words so that, together, they make sense and convey meaning. *Style* is reflected in the choice of words and the way we apply the rules of grammar in our writing. In traditional grade school curricula, grammar and style are the subject of much attention. What is important in technical writing, however, is not the ability to recite obscure grammatical rules; rather, it is being able to write correctly and effectively. In fact, grammar and style are important in technical writing for only two reasons:

1. Incorrect or improper grammar can change the meaning of what you are trying to say or, at least, make your meaning hard to decipher. That is fundamentally opposed to the goal of technical writing, which is precision in meaning.

2. Incorrect grammar says something about you and the quality of your thinking. Poor grammar in a technical report can communicate to the reader that you are not terribly bright or that you lack the required education or professional attention to

detail. Right or wrong, true or false, fair or unfair, poor grammar can seriously undermine your credibility.

In a technical document, you are judged to some extent based on the document's quality. By the time a reader is attempting to understand what you have written, you are not there to defend yourself, you are not available to explain what you really meant, and you have no opportunity to fix your errors.

Fortunately, most technical writers who make errors in grammar or style do so in relatively few areas. So this chapter focuses on the most frequent grammar and style mistakes in technical reports and will provide some straightforward solutions. Here are the most common problem areas:

- Comma splices
- Fused sentences
- Sentence fragments
- Misplaced modifiers
- Passive voice
- Verb agreement
- Pronoun agreement
- Pronoun references
- Case
- Spelling
- Noun clauses

Comma Splices

Comma splices occur when we join one sentence with another sentence by using a comma instead of a conjunction. This mistake is easy to make and easy to correct. Here is an example:

Comma splice

The circuit operates at dc, Ohm's law applies.

Here we have two sentences: "The circuit operates at dc" and "Ohm's law applies." The comma that follows "dc" is splicing these two sentences together, which is not something commas are supposed to do. To fix the problem, you have two choices: Since the two clauses are closely related, either you can join them with a semicolon, which functions as "soft" conjunction in this instance; or you can just use a comma and a conjunction.

Use a semicolon instead of that splicing comma.

The circuit operates at dc; Ohm's law applies.

Or follow that splicing comma with a conjunction.

The circuit operates at dc, and Ohm's law applies.

Fused Sentences

A *fused sentence* is a comma splice without the comma. In other words, two sentences are fused without any mark of punctuation:

Point of fusion

The workstation was not designed ergonomically it leaves much to be desired.

Again we have two sentences: "The workstation was not designed ergonomically" and "It leaves much to be desired." Note how they just run together at the indicated point of fusion. The solution, as with a comma splice, is easy:

Insert a semicolon.

The workstation was not designed ergonomically; it leaves much to be desired.

Or add a comma and a conjunction.

The workstation was not designed ergonomically, and it leaves much to be desired.

Or add a semicolon, an adverb, and a comma.

The workstation was not designed ergonomically; consequently, it leaves much to be desired.

Sentence Fragments

For a sentence to be complete, it must contain a verb. *Fragments* usually occur when the writer substitutes something else for this verb or leaves out the verb altogether:

Here *testing* is not a verb.

Tensile testing the specimen carefully with high levels of precision.

This is a fragment because it does not contain a real verb, and it does not make sense. The word *testing,* derived from the verb to *test,* does not act as a verb here; rather, it is a gerund, which is a verb used as a noun. To fix this problem, you need to add a verb:

Verb added here

Tensile testing the specimen carefully with high levels of precision is necessary.

Sentence fragments also occur when we put a subordinator before an otherwise perfectly good independent sentence:

Subordinator added here

Because the transformer could not take the load.

The subordinator *because* makes this clause dependent on something else, but the something else is not there. To fix the problem, you can remove the subordinator:

The transformer could not take the load.

Or you can add the "something else":

Independent clause added.

Because the transformer could not take the load, the system quickly failed.

Misplaced-Modifier Errors

English relies on word order or placement for meaning. In effective technical writing, *modifiers* have to be close to the words they are supposed to modify. Sentences in which modifiers are misplaced may be grammatically correct, but often they will not mean precisely what the writer intended. Here is an example of a misplaced modifier:

Ignorance of science is a phenomenon in society that must be destroyed.

Misplaced modifier

The writer is advocating destroying the ignorance of science in society. Because of the misplaced modifier, however, the writer is proposing that we destroy society. To fix this problem, move the modifier so that it relates more directly to what it is supposed to be modifying:

> Ignorance of science is a phenomenon <u>that must be destroyed</u> in society.

Passive Voice Problems

Passive voice and *active voice* refer to the movement of action through the sentence. In an active sentence, the subject comes first, next the verb, then the object of the verb's action. In a passive sentence, the object comes first, then the verb; the subject appears after the verb, if it shows up at all.

Consider the following active and passive sentences:

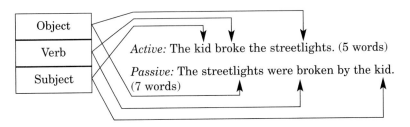

| Object |
| Verb |
| Subject |

Active: The kid broke the streetlights. (5 words)

Passive: The streetlights were broken by the kid. (7 words)

In the active sentence, the subject is the kid. The kid's action involved breaking, and the object of this action was the streetlights. The second sentence, the passive one, says exactly the same thing, but the object of the action comes first. The sentence is longer, and its construction is weaker. In addition, notice that the verb in the passive sentence has an auxiliary verb—a form of the verb to *be.* Instead of just *broke,* the verb now consists of the verb *were* and the past participle *broken.* Passive constructions always pair a form of the verb to *be* with the past participle of a verb. That is a good way to spot passive voice, although such pairings are not always passive and this method is not foolproof. The final test for passive voice is whether the object comes before the verb.

Passive constructions are weak, so why do people write in the passive voice? One reason is that it allows them to hide responsibility for their actions. Consider this modified passive voice sentence:

Modified passive: The streetlights were broken. (*no subject*)

Object

Verb

What is missing? The subject, of course: whoever broke the streetlights. The sentence may be grammatically correct, but it leaves out an important piece of information. We no longer know who is responsible!

People sometimes use the passive voice to hide their culpability in something that is bad; then, in the same sentence, they switch to active voice to take credit for something that is good. Consider this compound sentence:

Your medical records were lost, but I found them.

Here the passive voice in the first clause hides who lost the medical records, but the writer has switched to the active voice in the second clause to take credit for finding them.

Active voice is preferred in technical writing because it is more direct, it is clearer, and it provides the most information with the fewest words. However, passive voice does have its place in technical writing. It can be useful when the subject of the sentence is unimportant or obvious, or when the object of the sentence is the primary focus. Passive voice can also be a useful way of breaking the pattern of sentence structure to keep the reader from falling asleep. Finally, as mentioned, passive voice may be useful when one wants to hide responsibility.

Verb Agreement Errors

Verbs must *agree* with their subjects in person and number. If the subject is in the first person, the verb has to be a first-person verb. That is why "I is smart" is wrong and "I am smart" is correct. Many verbs in the English language are not person-sensitive, and for them, person does not matter.

In addition, if the subject is singular, its verb has to be singular. This requirement is the source of most verb agreement errors because, unlike agreement in person, some agreement-in-number errors do not sound wrong.

Here is an example:

Were does not agree with *implant.*

The implant, along with its associated circuits, were inserted into the patient's chest cavity.

The subject of this sentence is *implant,* which is singular. There is only one implant, but the verb *were* is plural. The verb does not agree with the subject, so a verb agreement error exists. This sentence may sound correct because the qualifier *along with its associated circuits* comes between the subject and the verb. Although *associated circuits* sound plural, the verb must agree with the subject, not the qualifier.

You can fix this sentence either by making the subject plural or by making the verb singular. Changing the subject, *implant,* to its plural form might be grammatically correct, but technically it is not accurate—unless, of course, the surgeon is really putting two or more implants into the patient's chest cavity. If we assume that only one implant is being inserted, the only option is to make the verb singular, as follows:

The implant, along with its associated circuits, was inserted into the patient's chest cavity.

Pronoun Agreement Errors

Pronouns must *agree* with their antecedents (the words they replace) in person, number, and gender. This rule is somewhat similar to the subject-verb agreement rule. Consider the following sentence, in which the pronoun does not agree with its antecedent in number:

Plural pronoun does not agree with singular antecedent.

Everyone in the lab must replace their radiation badges.

Believe it or not, *everyone* is singular (it comes from the two words *every* and *one*). The possessive pronoun *their* is plural, and trying to make it agree with *everyone* is a grammatical infraction. The solution is to make the antecedent plural or the pronoun singular. Consequently, either of these two sentences is correct:

Singular: Everyone in the lab must replace his or her radiation badge.

Plural: All people in the lab must replace their radiation badges.

One other point worth mentioning here involves political correctness. Notice that plural pronouns in the English language are not gender-specific. That means you do not have to worry about what gender they are. It is generally safer to avoid specific gender references by using plural nouns and pronouns. *Their* is not gender-specific and, consequently, represents a safe approach—as long as the antecedent includes more than one person. If the antecedent refers only to one individual, use that individual's gender; if you do not know the person's gender, then use *his or her* or *her or his.*

Pronoun Reference Errors

The other common pronoun error in technical reports involves the use of a pronoun whose antecedent is unknown or unclear. In technical writing, pronouns must refer clearly and without question to specific antecedents. Here is an example:

> The coolant leak impaired the CPU's heat dissipation, resulting in an erroneous reading at the most critical part of the process. This had a cascading effect on the system.

To what does *this* refer? It could refer to the coolant leak, to the erroneous reading, or to both. Such ambiguity can be problematic in technical writing, especially when you are using the pronoun *this.* In fact, a good rule is to always include a noun after the pronoun *this.* You could fix this sentence as follows:

> The coolant leak impaired the CPU's heat dissipation, resulting in an erroneous reading at the most critical point in the process. This coolant leak had a cascading effect on the system.

Case Errors

Case errors involve putting a noun or pronoun in the wrong case. The three English cases are subjective, objective, and possessive. The *subjective* case (also called the *nominative* case) is what we put subjects in. The *objective* case is what we put objects in. The *possessive* case shows possession. Here is an example:

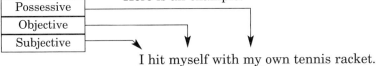

Possessive
Objective
Subjective

I hit myself with my own tennis racket.

The first pronoun, *I,* is in the subjective case because it is the subject of the sentence. The second pronoun, *myself,* is in the objective case because it is receiving the action (*myself* is actually the reflexive form of *me*). The third pronoun, *my,* is in the possessive case because it shows possession of the tennis racket. In "I hit myself with me own tennis racket," the *me* represents a case error. (It would also sound ridiculous—unless, of course, I were Scotty in the old *Star Trek* series.)

The cases of pronouns such as *who* and *whom* are a cause of frequent errors. *Who* is subjective (Who stole my watch?), while *whom* is objective (From whom was my watch stolen?). Case errors can also get a little tricky in some instances:

Subject of a gerund ——┐   Gerund
                              ↓   ↓
The transmission microscope malfunctioning caused the experiment to be delayed.

In this example, the word *malfunctioning* is a gerund, or a verb used as a noun. A rule about gerunds says that their subjects are always in the possessive case. The subject of this gerund is *microscope,* which should be possessive in this sentence:

The transmission microscope's malfunctioning caused the experiment to be delayed.

A better approach would be to put the gerund first as the subject of the sentence.

The malfunctioning of the transmission microscope caused the experiment to be delayed.

Spelling Errors

Most people assume that an educated professional can spell words correctly. Spelling errors make your reader doubt the validity of your writing. So check and recheck your writing for spelling errors.

Consider how these errors would look in a technical paper:

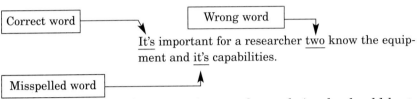

The expression *two know* obviously should be *to know,* and *it's* should be *its* (no apostrophe for the possessive pronoun that should be used here). The correct sentence reads as follows:

> It's important for a researcher to know the equipment and its capabilities.

A Word about Homonyms

The errors in the sentence above all involve homonyms, which are words that sound alike but have different meanings. Substituting *to* for *two* or *it's* for *its,* as was done in the sentence above, is a homonym error that computer spell checkers rarely catch, but which can look really stupid in a technical report. The problem is that the English language has many homonyms that lend themselves to these kinds of errors. How many writers have used *bare* (meaning naked) instead of *bear* (the big furry animal), or *oral* (something involving the mouth) instead of *aural* (something you hear)? Many technical writers also routinely confuse words that sound almost alike. For example, while not technically homonyms because they sound slightly different, consider the words *effect* and *affect. Effect* is the noun meaning *the result of some cause,* or a verb meaning *to cause something to happen;* and *affect* is a verb meaning *to have an influence on something.*

Keeping homonyms straight can be challenging. Just consider the editor who marked *carets* on her proofreading copy, who wore a 2-*carat* diamond in a ring setting made out of 18-*karat* gold, and who ate a *carrot* every day in her dinner salad. The bottom line is: Stay alert for homonym errors!

Spelling and Numbers

An additional spelling consideration worth mentioning here involves numbers. One of the most common errors is to start sentences with a numeral:

<u>9,192,631,770</u> hertz is the spectral line frequency of cesium 133.

Correcting this kind of error can be difficult, but you have to get rid of that leading number because it is unattractive and can be confusing in the middle of a document. The *worst* fix would involve spelling out the number:

<u>Nine billion, one hundred ninety-two million, six hundred thirty-one thousand, seven hundred and seventy hertz</u> is the spectral line frequency of cesium 133.

An easier solution would be to simply insert an appropriate word before the number:

<u>Exactly</u> 9,192,631,770 hertz is the spectral line frequency of cesium 133.

Or simply invert the sentence:

The spectral line frequency of cesium 133 is 9,192,631,770 hertz.

When smaller numbers start a sentence, you can just spell them out:

Thirty-two degrees Fahrenheit is the freezing point of water.

Finally, pay attention to which words you capitalize. Normally, in technical writing, some type of convention or style sheet will tell you what you should capitalize. Many style guides exist, along with lots of differences regarding the use of capital letters. Always follow the style guidelines for your organization or activity.

A good general approach is to avoid unnecessary capital letters. Consequently, if you are going to capitalize a word, have a specific reason for doing so. Here are some common reasons for capitalizing words:

- Capitalize names of specific persons, places, or things (proper nouns). These include the names of specific people (Albert Einstein); cities or street names (Dayton, Ohio; First Street); historical documents and religions (the Declaration of Independence, Catholicism); titles of papers, books, films, software packages, and trademarks (*Gone with the Wind,* Microsoft Word); and periods of time or geographic regions (Cambrian Period, the West).
- Capitalize abbreviations or acronyms (ATM, Ph.D.).
- Capitalize titles that precede a person's name, but not those that follow (Professor John Smith, but John Smith, the professor).
- Capitalize the first word of every sentence and the pronoun *I.*

Noun Clauses

You can think of a noun clause as a short sentence used in its entirety as a noun in a longer sentence. These constructions can be troublesome for technical writers because the individual elements of a noun clause are treated as elements of a complete sentence, while the entirety of a noun clause is treated as a single element in another sentence. Confusing? You bet it is. Consider the following two versions of the same sentence:

> *Version 1:* The boss decided <u>whom</u> will go to the conference.

> *Version 2:* The boss decided <u>who</u> will go to the conference.

At first glance, it might seem that version 1 with *whom* is correct, since *whom* is the object of the verb *decided* and should be in the objective case. But it is not that simple. The pronoun *whom* is *not* the object of *decided*. It is only part of the object. The entire clause that follows *decided* is, in fact, the object. Since *whom* is the subject of that clause, it needs to be in the subjective case, as in version 2, and should be *who*. Also, since the entire clause *who will go to the conference* acts as the object of the verb *decided,* the entire clause functions as a noun and is therefore considered to be a *noun clause*.

Stylistic Considerations

Style is a general concept that refers to the way we say something. Style encompasses many things, including word choice, order, and selection; and it actually includes any other distinctive features of the way we express ourselves in language. Great! So what does style have to do with technical writing? Besides being grammatically correct, technical writing must be geared to communicating precise information in a straightforward and unambiguous way. It is easy to be ambiguous and imprecise while still being grammatically correct. For technical writing to be effective, however, it must use a style of expression that is economical and precise.

Economy

One of the most important things you can do in technical writing is to say what you want to say without making your reader endure a "bunch of fluff" in the process. Consider the following sentence from a technical proposal cost section.

> With the availability of all necessary equipment, along with required materials and a student who can perform required tasks at the level of competency appropriate for the tasks involved, we believe that all satisfactory and unsatisfactory options can be fully and completely considered, and a reasoned and effective recommendation provided within the time and cost constraints of the proposal, given the past, present, and future uses of the technology.

Look at the style of this sentence. What a disservice to the reader! What about *all satisfactory and unsatisfactory options*—as opposed to what other kinds of options? No other kinds of options exist! How about a *reasoned and effective recommendation,* as opposed to what—an unreasoned and ineffective recommendation? And what about *past, present, and future uses?* That pretty much includes everything, does it not?

This kind of writing is ridiculous. Much of this paragraph is pure padding, and most of the words add absolutely nothing to the meaning. In fact, this kind of writing actually confuses or dilutes the point the author is trying to make. We could easily rewrite this paragraph and say the same thing in a much more straightforward way:

> With the required equipment and materials, and a competent student, we can fully consider all options and make a recommendation within the constraints of the proposal.

Get to the point as quickly as possible. Look at the following example, a letter from higher management sent to the technical staff of a company:

To All Technical Staff:

Centuries ago, a generalized philosophy of the earth and the cosmos constituted scientific thought and did not require exactness. The very essence of science and technology today relies on a foundation of reproducible standards, and since 1866, this country has specified the International System of Units (SI), a.k.a. the *metric system,* as the only legal system for electricity and illumination. Standardized nomenclature is frequently victimized by the conflicting approaches of SI and the English system, even though SI provides the standard prefixes required for dealing with a wide range of units. Consequently, all technical staff will use only SI in company documentation.

Of course, we could shorten this paragraph by just getting to the point:

To All Technical Staff:

Use the metric system in all company documentation.

There is no need to "soften up" the employees with an esoteric and irrelevant discussion of SI. It is unlikely that the company's technical staff will fall to the ground in despair when told up front that they should use the metric system.

Precision

Because technical writing needs to be precise, we need to select words that are precise in meaning. Look at this example:

The X-49 is a very large chip that costs a lot of money. It has unprecedented power; in fact, its predecessor, the X-45, is just a faint shadow of the X-49.

So, how big is *very large?* Is the X-49 larger than the solar system? That certainly would be *very large* unless, of course, we were comparing the solar system to the Milky Way galaxy. Then it would be *really small.* And what exactly does *unprecedented power* mean? Is that more than a *whole lot of power?* Also, if the X-45 is just a *faint shadow* of the X-49, does that mean it is *a lot smaller,* or maybe just *somewhat smaller?* You get the idea. Avoid words such as *very, awfully, extremely, really, a lot,* and *kind of* in technical writing. Such words may sound good in some cases, but in reality, they have no precise meaning.

The sentence above could be rewritten precisely as follows:

The X-49 is an Ardvark1000-socketed chip that costs $9650 per unit to manufacture. Its performance benchmarks are two orders of magnitude greater than those of its predecessor, the X-45.

CHAPTER FOURTEEN

Documentation

Your technical writing teacher has just assigned a large research report that must have at least 10 sources, not including encyclopedias or websites. To save time and effort, you obtain everything you need from (where else?) an encyclopedia or website. Then, using that information, you whip up a complete research report. Next, through quick Web searches and a superficial skimming of newspapers, magazines, and books, you identify multiple sources that somehow relate, even tangentially, to the topics about which you have already written. These become the "sources" that you reference in your paper to keep the teacher happy.

The problem with this approach—besides leading to a paper with inferior content and throwing all standards of individual integrity and academic ethics into the dumpster (see ethics in Chapter 2)—is that it compromises the role of documentation in technical writing. It ignores the need for complete, accurate source citations in the most serious form of business and scientific writing.

Technical writing often involves big business, lots of money, and a competitive, unforgiving environment. Proper documentation is an essential element of technical writing—an element that can have serious legal, ethical, and credibility implications for those who fall short of the mark. Documentation requirements are something that any technical writer must take seriously.

What Is Documentation? In its most general meaning, *documentation* refers to creating virtually anything recorded or "documented" on paper. All the documents described in this book are forms of technical documentation. However, this chapter focuses on the kind of documentation that involves referencing sources. In this sense of the word, documentation gives formal credit to a person, organization, or publication for an idea or information that either is not original or is not common knowledge of the field. It represents an acknowledgment of your indebtedness to the source.

For example, if you need to quote the performance specifications for the 16XL1000000 transmitting tube from a *Boss*RF technical report, you must document that technical report. Those specifications represent ideas that are not original—in other words, they are not your ideas. On the other hand, if you use Ohm's law to show that 10 volts across a 1-ohm load produces 10 amperes of current, you do not need to document your source. Even though $I = E/R$ is not your idea, it is common knowledge in the field of electronics and, as such, does not need to be referenced.

Whether something is common knowledge of the field is often a judgment call. Normally, we think of something as being common knowledge when the average skilled person in the field should already be familiar with it. However, the best approach is to document any source when you are in doubt. Not only does this ensure that those who deserve credit receive it, but also it usually enhances the credibility of your writing by adding some authority to the argument.

Your goal in providing documentation should be to give enough information about the sources you have used to enable your reader to find and consult those sources conveniently and independently. Sources typically include print media such as books, journals, periodicals, newspapers, conference proceedings, and dissertations; electronic media such as websites, file transfer protocol (FTP) servers, newsgroups, and forums; storage media such as CDs and DVDs; and other material such as lectures and interviews.

Many style guides exist today for documenting sources. Here are a few: *MLA Handbook for Writers of Research Papers,*[1] *The Chicago Manual of Style,*[2] the *Government Printing Office Style Manual,*[3] the *APA Publications Manual,*[4] the American Chemical Society's *Manual for Authors and Editors,*[5] the *American Institute of Physics Style Manual,*[6] and *The Council of Biology Editors Manual for Editors and Publishers.*[7] The best approach is to use the style guide specified by your employer, your teacher, or your field. If a style guide is not specified, you can use any consistent form of documentation that provides the necessary information.

Documentation Styles

As a technical writer, you should document sources for any or all of the following reasons.

When to Document Sources

To Meet Legal Requirements

Legally, when using copyrighted sources, you are required to document these sources. Federal copyright statutes control the reproduction of original works, including books, music, drama, computer programs, databases, videos, sculptures, and virtually any other media. Although copyright laws do not specifically protect the ideas contained in these works, they do protect the expression of ideas by these works. Also, some works, such as photographs, contain ideas that are intertwined with expression and cannot be separated easily.

You may use copyrighted material, either with permission of the copyright holder or without permission if your use is within the scope of the *Fair Use Doctrine*. *Fair use* provides for limited reproduction of copyrighted material without the permission of the owner for noncommercial, teaching, and research purposes. Fair use also requires that the original work be fully documented (referenced). If you do not provide this documentation, then it is not fair use—it is a violation of copyright law.[8]

To Meet Academic Standards

Academic standards require that you document any nonoriginal ideas, except those that represent common knowledge of the field. You must document not only direct and indirect quotations, but also paraphrases or any other discussions that specifically refer to or include original ideas that are not yours.

To Establish Credibility

You should support your original assertions or conclusions, which are not based on common knowledge, when you can show complementing positions on the part of authoritative sources. The purpose here is to establish the credibility of your position. Avoid making unsupported assertions in technical documents. If your assertions are consistent with the ideas of a recognized authority or previous work, document (reference) that authority or work to establish your credibility.

How to Document Sources

Generally speaking, two different approaches exist for documenting: *notational* and *parenthetical.*

- *Notational documentation* places footnote or endnote superscript numbers in your paper at the point where you need to document a source. You include the actual source citation either as a footnote at the bottom of the page or as an endnote at the conclusion of the report (or a subsection of the report).

- *Parenthetical documentation* places a source citation in parentheses in your paper at the point where you need to document a source. You include a list of references at the end of the report (or major subsection). The citations are keyed to that list of references by either number or author's last name.

Most technical documents use parenthetical documentation because it is simple and effective. Parenthetical documentation also gives the reader more information at the point of the citation.

Parenthetical documentation consists of two parts: the list of references at the end of the report or subsection and the parenthetical references cited within the text of the paper.

- The *list of references* (also called *list of sources, sources, references, notes, works cited,* or, frankly, whatever sounds reasonable) is essentially a bibliography that provides specific information (author, title, publication, date) about the works used or considered by the writer. You can list the references alphabetically by the first significant word, which is usually the author's last name. Or you can sequentially number your sources. If you list your references alphabetically, you usually can reference the source with the author's last name. Numbering your sources will make things even easier because you can reference each source in the paper by its number in the list.
- The *parenthetical references* are inserted into the text of the paper as the source citations. In the parentheses, first you identify the source by either source number or author's last name along with the date of publication, and then you add the specific pages being referenced (if applicable—some

sources do not have page numbers). For example, consider the following paragraph:

> The 16XL1000000 is a high-power transmitting tube normally used in class C radio-frequency applications. However, as research commissioned by the Village Thumpers so clearly demonstrates, the tube can be used successfully as a power amplifier in class A audio applications as well. Its power, distortion, and signal-to-noise performance specifications are quite impressive in this regard. (3, pp. 22–23) or (Yinburg 2005, pp. 22–23)

This particular paragraph needs a reference to the Village Thumpers' commissioned research. Either of the two sample citations will work. The first example uses a source number, whereas the second example uses the source's last name—in this case Dr. Robert W. Yinburg's last name because he is Thumper Enterprises' chief scientist and wrote the referenced report.

As mentioned, the sources themselves are listed at the end of the report. This list of references might look something like the following:

List of References

1. Pradeep Misra, "Order Recursive Gaussian Elimination," *IEEE Transactions on Aerospace and Electronic Systems,* AES-32 (January 1996), pp. 396–401.
2. T. Sudkamp, *Languages and Machines: An Introduction to the Theory of Computer Science,* 2d ed. (New York: Addison-Wesley, 1996).
3. Robert W. Yinburg, "Preliminary Report on the 16XL1000000 in Audio Service," TR-05-0022. Thumper Station, N.Y.: Thumper Enterprises, September 2005.

The parenthetical documentation in our example was either (3, pp. 22–23) or (Yinburg 2005, pp. 22–23). Notice that both point to the same source in our list of references. Whichever parenthetical format you decide to use, be consistent throughout the entire report. Do not list a source by number in one section and by author's last name in another.

You may also want to add sources that you consulted but did not specifically use. The idea is not to pad your reference list, but to acknowledge sources that may have influenced your thinking even though you did not specifically cite them. These latter sources should be included in a separate list of references clearly indicating that they were consulted but not used. In other words, you might end up with a "List of Sources Consulted and Used" and a "List of Sources Consulted but Not Used"—or perhaps "Sources Cited" and "Sources Consulted."

Again, it is best to use the style guide specified by your boss, your teacher, or your field. However, if you do not have a specified format, the simple approach provided in this chapter should be adequate for documenting the most commonly used sources in technical writing.

Of course, several different types of sources exist, and each type is handled differently in the list of references. The following examples will show you how to handle the most common types of sources.

The following examples provide a general guide for documenting the most commonly used forms of print media.

Print Media Examples

Books

Include the author(s), title, edition, city of publication, publisher, and date of publication:

> **1.** T. Sudkamp, *Languages and Machines: An Introduction to the Theory of Computer Science,* 2d ed. New York: Addison-Wesley, 1996.

Journals

Include the author(s), article title, journal, volume (and number if necessary), date, and inclusive pages:

> **2.** M. E. Brown and J. J. Gallimore, "Visualizing of 3-D Structures During Computer-Aided Design," *International Journal of Human-Computer Interaction,* 7 (1995), pp. 37–56.

Conference Papers

Include the author(s), paper title, conference or transactions information, date, and inclusive pages:

> **3.** R. V. Grandhi and L. Wang, "High-Order Failure Probability Calculation Using Nonlinear Approximations," 37th SDM Conference, Salt Lake City, Utah, April 1996, pp. 1292–1306.

Encyclopedias

Include article, encyclopedia, edition, place of publication, publisher, date, and inclusive pages:

> **4.** "Acceleration," *Von Nostrand's Scientific Encyclopedia,* 9th ed., New York: Wiley-Interscience, 2002, p. 10.

Newspapers

Include author(s) if known, article title, newspaper name, date, section, and page(s):

> **5.** George Will, "Volcano's Reverberations Still Felt?" *Dayton Daily News,* May 22, 2003, sec. A, p. 15.

Note: In this case, the author is listed. If no author were listed, you would alphabetize the entry by the first significant word of the article title.

Nonjournal Entries

Include author, article title, publication, date, and inclusive pages:

6. Harry Goldstein, "Irradiation Nation," *IEEE Spectrum,* August 2003, pp. 24–29.

Technical Reports

Include author, title, number, agency, place, and date:

7. Fred D. Garber, "Synthetic Aperture Radar Automatic Strategic Relocatable Target Identification System," WL-TR-93-1145, Wright Laboratory, Wright-Patterson Air Force Base, Ohio, October 1993.

Dissertations and Theses

Include author, title, degree level, school, and date:

8. Albert J. Rosa, "Luminescent and Electrical Properties of Sodium Implanted Zinc Selenide," Ph.D. dissertation, University of Illinois at Urbana-Champaign, 1975.

The following examples provide a general guide for documenting electronic media. Documenting electronic media, especially sources that exist solely in cyberspace, represents significant challenges for the technical writer. Sources range from specialized databases (e.g., Lexis-Nexis), to online service databases (e.g., AOL), to forums and news groups (e.g., USENET), and to the mass of material now available to anyone on millions of Internet websites.

When you document cyberspace sources, keep in mind the differences between cyberspace and traditional sources. For example, an Internet address

Electronic Media Examples

does more than just indicate where a source is located. It also provides the detailed electronic directions, including the precise path, for accessing the source. In effect, a cyberspace address not only shows you where the source is located, but also takes you there!

Unfortunately, computer networks are absolutely unforgiving of errors in addresses. If you make even the slightest mistake in order, spelling, or punctuation, you will not reach your desired Internet site. To make matters worse, long-held use of certain marks of punctuation such as periods and slashes is often incompatible with Internet addresses. For example, you may not be able to use a period at the end of a sentence when working with Web addresses. Consider this sentence:

> For more information on Dr. Finkelstein's courses, go to http://www.finkelnet.com.

If you are Web-savvy and do not enter the trailing period, you will reach the author's finkelnet.com Web page, which is used to support his classes. If you enter the final period, however, the finkelnet .com address will fail, resulting in a "server not found" error. In hypertext markup language (HTML), the period's function has nothing to do with English grammar and punctuation. Consequently, you would have to add a space before the ending period (the terminal mark of punctuation), or invert the sentence to move the Web address away from the end of the sentence, or even omit the final period altogether.

Another problem with cyberspace documents is the lack of key information. The goal when you are documenting an electronic source in "a perfect world" is to provide the reader with all the relevant information about that source. But in cyberspace, the world is far from perfect. For example, the author's name may not be included on the

Web page, and the title may be missing or only included in a /title tag in the HTML code. Additionally, the universal resource locator (URL) or Web address may be redirected from one page to another without your knowledge or any obvious indication. You may not be getting the information from where you think.

Finally, cyberspace addresses tend to be unreliable over time. Websites come and go, and material on websites is "here today and gone tomorrow." On some sites, Web addresses are dynamically assigned to time-sensitive materials, then reassigned once the materials are archived. In other words, an address that works today may not work tomorrow—or even later today! Generally, the best advice when you are documenting a cyberspace source is to do the best you can do with the information available. Remember, your goal in documentation is to provide your reader with enough information so that he or she can independently locate the source. When all else fails, even providing partial information—perhaps a topic or an author's name—might be enough with modern search engines to track down the source. When documenting electronic sources, however, you are obligated, to the maximum extent possible, to include all the relevant information. In this case, the information would include the author, title, database record identifier (file name and path), medium, and date of posting (or date of access, if the posting date is unknown).

Here are a few examples.

Website

Include author (if known), title, medium, site owner, Web URL, and date:

9. "Tax Stats at a Glance." Internet: Internal Revenue Service, www.irs.ustreas.gov/taxstats/article/0,,id=102886,00.html, accessed June 1, 2003.

In this example, the author is not known, so the reference begins with the title.

Online Forum

Include author (or topic), title, medium, site owner, complete network address, and date:

> **10.** Figure Skating, "ISU Press Conference Transcript of April 1, 2003." Internet: google.com: groups.google.com/groups?hl=en&lr=&ie=UTF-8&group=rec.skate, accessed June 2, 2003.

FTP Site

Include author, title, file name, medium, site owner, complete network address, and access date:

> **11.** Leo Finkelstein, Jr., "College Recruiting Game" (file=ecsslots.exe). Internet: College of Engineering and Computer Science, Wright State University, Dayton, Ohio, ftp.cs.wright.edu/~lfinkel, June 1998.

Computer Local Storage Media (Computer Disk, Flash Card, etc.)

Include file name, medium title, medium type, version, series or ID number, and date:

> **12.** "Readme," *Apple Hardware Test—iMac*, CD, version 2.0, 691-4199A, 2003.

Other Examples | **Interview**

Interviews include most situations where you pose questions to a source and receive answers. Interviews do not have to occur face to face or in real time. Include interviewee, method, topic, affiliation, place, and date:

> **13.** Leo Finkelstein, Jr., Personal interview, Topic: "The Ethics of Using the FinkelKICK for Self-Defense." Office of the Dean, College of Engineering

and Computer Science, Wright State University, Dayton, Ohio, April 21, 2002.

Lecture

Include lecturer, occasion, topic, location, and date:

14. Leo Finkelstein, Jr., EGR 335 class lecture, Topic: "Documentation." College of Engineering and Computer Science, Wright State University, Dayton, Ohio, May 8, 2004.

• Have I used source citations throughout the text keyed to my list of references?

• Have I documented all uses of copyrighted material?

• Have I documented all nonoriginal ideas that are not common knowledge?

• Have I referenced authoritative sources that support assertions I have made that are not otherwise supported?

• Have I used the prescribed method and form of documentation (if applicable)?

• Have I been consistent in the method and form of documentation I have used?

Checklist for Documentation

---

1. *MLA Handbook for Writers of Research Papers,* 5th ed. New York: The Modern Language Association, 1999.
2. *The Chicago Manual of Style,* 15th ed. Chicago: University of Chicago Press, 2003.
3. *Style Manual,* rev. ed. Washington, D.C.: U.S. Government Printing Office, 1984.
4. *Publication Manual of the American Psychological Association,* 5th ed. Washington, D.C.: American Psychological Association, 2001.
5. *The ACS Style Guide: A Manual for Authors and Editors.* Washington, D.C.: American Chemical Society, 1986.

Notes

6. *AIP Style Manual,* 4th ed. New York: American Institute of Physics, 1990.

7. *Scientific Style and Format: The CBE Manual for Editors and Publishers,* 6th ed. New York: Cambridge University, 1994.

8. For more information, see "About Copyright," Internet: the United States Copyright Office www.loc.gov/copyright/ September 2000.

CHAPTER FIFTEEN

Visuals

Remember the last time you went to the dentist? As you sat in the waiting room, you probably picked up one of the magazines lying on the table. What did you look at? In all likelihood, you looked at the pictures. In fact, if you read anything, it was probably the captions under the pictures because of the large amount of information provided.

What Are Visuals?

Visuals are presentations of ideas that exploit our sense of sight to communicate a large amount of data quickly and efficiently. You looked at the pictures first because of their information bandwidth; you can get more information from them than from reading the text, especially with the limited time available in the waiting room. You also looked at the pictures first because of their greater interest and higher credibility.

Technical writing deals with complex topics in precise ways. It is not surprising that one of the most important tools for a technical writer is visuals—things that we call figures, diagrams, drawings, illustrations, graphs, charts, schematics, maps, photographs, and tables. Whether you are showing an exploded view of a mechanism or plotting the regression curve from an experiment, you will find that visuals are an absolutely essential element of any technical report.

General Guidelines for Using Visuals

- Include visuals in a technical paper only when you have a reason to do so. If you do not know why you are putting a visual into a paper, you probably do not need it.

- Be sure to reference a visual in the text discussion prior to its placement in the report. If the visual precedes its reference, the reader will wonder why it is there.
- Explain the significance of all visuals, supplying your interpretation of the data depicted.
- Be sure to number and title all visuals.
- Ensure all visuals directly clarify or otherwise enhance the text discussion. You need to integrate them into your report, not just stick them somewhere. That means the labels and captions used in a visual should match the text discussion that refers to the visual. For example, if you are describing the negative terminal of a diode, do not call it the *negative terminal* in the text and the *cathode* in the visual.
- Document your visuals when they contain copyrighted information or represent borrowed ideas. Because visuals often get separated from the report, do not rely solely on notational or parenthetical documentation; include a source line with the visual itself, usually under the visual's number and title.

Guidelines for Design of Visuals

Reproducibility

Design your visuals with the output process of your report in mind. If your report will be printed or duplicated in a single-color ink or toner, consider that fact when you are developing graphs and diagrams. Be especially wary of different colors that may look great on your video screen but could print with exactly the same shade of gray or even blend into the color of the paper on which the report is printed. A safe approach is to use pattern fills instead of colors when the report will be printed or duplicated in a single color.

Simplicity

Remember that the purpose of visuals is to supplement and clarify the information you are presenting. Some concepts (such as precisely modeling the entire supply and distribution process of a large retail organization) are so complex that they do not lend themselves to visual presentation. Other concepts may need to be broken down into smaller components for effective presentation. For example, to show the operation of a laser printer, you would not include the charging electrode, electrostatic plate, scanning laser, toner reservoir, transfer roller, fusion roller, cleaning pad, and paper transport mechanism all in the same visual. Separate visuals for each part might be needed.

Accuracy

Ensure that any visual you use accurately portrays the information being presented. Do not exaggerate the data by manipulating scales or misrepresenting relative sizes. However, in certain documents, where it is clear that you are presenting a biased point of view, some visual enhancement of information is acceptable. For example, in a proposal, where you are selling your skills or your organization's ability to do a particular task, it is expected that you will enhance your strengths and downplay your weaknesses. That does not mean it is acceptable to lie about your strengths or weaknesses.

Types of Visuals

Visuals generally fall into one of the following categories: diagrams, graphs, schematics, tables, or images. The following sections provide a brief discussion of each.

Diagrams

Diagrams are drawings that show the components of a mechanism, the steps of a process, or the relationship among parts of a system. Make diagrams only as complex as they need to be. Diagrams can provide normal, cutaway, or exploded views of a mechanism. Figure 15.1 provides a normal, external view of the Quantum Central Processing Unit (QCPU) discussed in Chapter 11. This view would be useful for describing the external physical attributes of the device.

Figure 15.2 provides a two-dimensional, cutaway view of the same device. This view would be useful for describing the internal parts of the QCPU and perhaps the process of its operation.

Figure 15.3 provides a 3-D, exploded diagram of the same device. This view might be useful for describing the physical attributes of the internal structure, providing assembly instructions, or describing the process of its operation.

Graphs

Graphs are visual representations either of relationships among sets of numbers or of quantities and proportions of mathematical values. Graphs are great for presenting statistical information. Generally, the three types of graphs you will work with are line charts, bar and column charts, and pie charts.

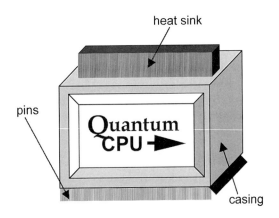

Figure 15.1
QCPU external view.

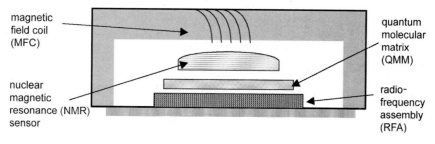

Figure 15.2
QCPU 2-D cutaway.

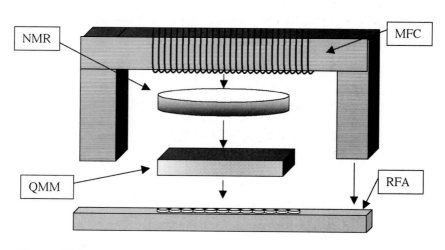

Figure 15.3
QCPU exploded 3-D view.

Line Charts

Line charts effectively show trends in data. Normally, the vertical y axis is used to plot dependent (variable) data points, and the horizontal (x) axis labels the independent variables. Each line plotted then shows the changing value of a specific variable. Figure 15.4 provides a line chart showing the 6-year stock value trends for Quantum Processors, Inc.

Bar and Column Charts

Bar and column charts provide excellent tools for comparing discrete variables. Since each variable is a separate entity, a line chart would not be appropriate because no slope actually exists between data points. Normally, you use the bar chart (Figure 15.5) when the value labels are too long to conveniently fit on the horizontal axis. You use the column chart (Figure 15.6) when the value labels are short enough to fit well on the horizontal axis. Note that both of these figures provide discrete (and, unfortunately, fictitious) data regarding the author's income sources.

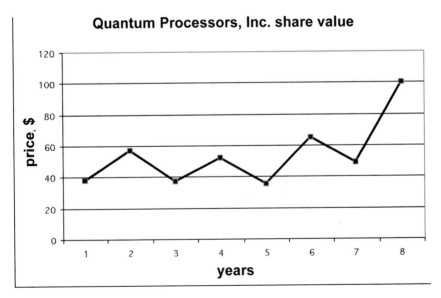

Figure 15.4
Line chart of critical point value trends. *(Source: Quantum CPU International, Technical Report 002345-S, August 2004, p. 21.)*

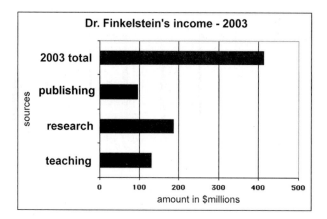

Figure 15.5
Bar chart of Finkelstein's income. (*Source:* "The Kind of Guy Your Daughter Should Marry," *Affluent American Magazine,* February 2003, p. 41.)

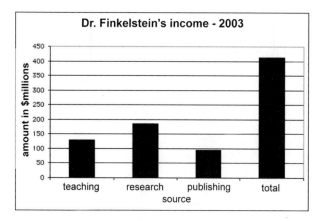

Figure 15.6
Column chart of Finkelstein's income. (*Source:* "The Kind of Guy Your Daughter Should Marry," *Affluent American Magazine,* February 2003, p. 41.)

Pie Charts

Pie charts are useful for showing the relative proportions of a whole that each discrete data category represents (Figure 15.7). Pie charts, however, do not provide the same degree of visual precision as line, bar, or column charts.

Adding Visual Interest to Line, Bar, Column, and Pie Charts

In some situations, you might need to add visual interest to a chart to get your reader to look at it. A simple way to do that is to add a graphic or photograph to the chart. Another way is to use 3-D line, bar, column, and pie charts. That extra dimension of depth will add visual interest to the charts, but it can also make the charts less precise. If your goal, however, is to create dynamic impact, 3-D charts might work well. Pie charts can also benefit from such presentation, especially because they are not used for precision anyway.

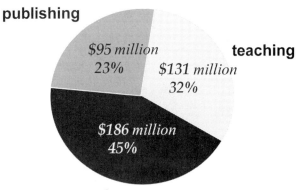

publishing

teaching

research

$95 million 23% *$131 million 32%* *$186 million 45%*

Figure 15.7
Pie chart of Finkelstein's income. (*Source:* "The Kind of Guy Your Daughter Should Marry," *Affluent American Magazine,* February 2003, p. 41.)

The most dramatic way to treat a technical chart is to turn it into a pictograph. Figure 15.8 is a pictographic chart of the author's income (in his dreams). This particular chart is analogous to the bar chart in Figure 15.5 and the column chart in Figure 15.6. Instead of bars or columns, this chart uses piles of coins as analogs for the values of each category. Obviously, this chart lacks the precision of either the bars or columns, but it is far more interesting.

Schematics

Schematics visually represent a system's structure or the procedures involved in a process. For example, the flowchart in Figure 15.9 schematizes (and oversimplifies) the writing process that one might use to produce a report.

Schematics are also used frequently to provide circuit diagrams, such as the one of the power amplifier circuit in Figure 15.10.

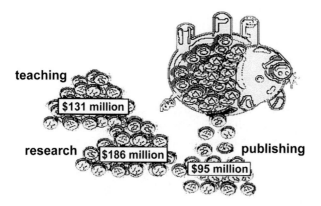

Figure 15.8
Pictographic of Finkelstein's income. (*Source:* "The Kind of Guy Your Daughter Should Marry," *Affluent American Magazine,* February 2003, p. 41.)

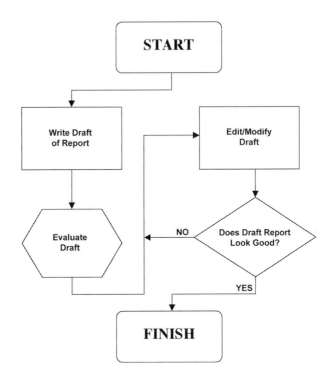

Figure 15.9
Schematic of report writing process.

Tables

Tables are orderly arrangements of data and information in columns and rows. Tables are the most precise way to display an array of data. They also can be used for a wide range of technical writing applications, from simple listings to complex troubleshooting charts. Table 15.1 provides, in tabular form, the precise income data on which Figures 15.5 through 15.8 were based. Table 15.2 shows a troubleshooting chart for setting up wireless connectivity on a notebook computer.

Table 15.3 shows a simple listing of Finkelstein's purchases. Figure 15.11 is a pictographic representation of the same data.

Table 15.1 Dr. Finkelstein's income
for 2003

Source	Amount
Teaching	$131,002,130
Research	186,000,456
Publications	95,000,101

Source: "The Kind of Guy Your Daughter
Should Marry," *Affluent American Magazine,*
February 2003, p. 41.

Table 15.2 WiFi 802.11b troubleshooting chart

Problem	Causes and Remedies
Computer does not see base station	Wireless device not installed *Check device manager for status* Wireless device on wrong channel *Reset channel higher or lower*
Base station signal is very weak	Wireless device antenna not enabled *Enable the antenna* Location is poor *Move to a different location*
Computer not allowed access	WEP code or ID missing or wrong *Check and re-enter code or ID*

Table 15.3 Tabular listing of Finkelstein's
purchases

Dr. Finkelstein: What does he buy?	
Toys	Friends
Food	Computers
Drink	Aphrodisiacs
Books	Videos
Travel	Home repair
Utilities	Happiness

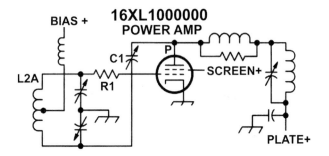

Figure 15.10
Power amplifier circuit.

Figure 15.11
Pictographic of Finkelstein's purchases.

Photographs

Photographs are visual reproductions, recorded on film emulsion or encoded as digital data, that accurately reproduce the appearance of objects or events. They can add interest, credibility, and extreme visual detail to a technical report. Photographs are high-bandwidth tools that can give your reader lots of credible, nonverbal information quickly.

For example, consider how photographs are used in the following brief discussion of lava flows.

Lava is a type of igneous rock created when molten rock called magma reaches the surface through volcanic action. Certain volcanoes, known as *shield volcanoes,* produce large amounts of hot molten rock called magma (2100°F). Magma pouring from the volcano often creates expansive flows of hardened lava. One of the most active shield volcanoes with extensive lava flows is Kilauea on the island of Hawaii.

Kilauea tends to produce two main types of lava: *Pahoehoe* and *'a' a.* Pahoehoe lava is formed when a relatively shallow flow (1 to 2 feet thick) of magma cools and starts to harden. The crust that is formed in Pahoehoe lava can have a smooth, glassy surface; or it can have a ropelike, wrinkled texture caused when the hardening surface, which cools first, is dragged and pushed by the molten lava below. On the other hand, 'a'a lava flows can be several yards thick and are characterized by rubble that has broken off from the flow's leading edge, forming a cascade of debris (see Photo 15.1).

The photograph adds a dimension to this discussion that allows the reader to better understand and appreciate the nature of lava flows. For one thing, looking at the photograph better enables the reader to appreciate the size and

Insert: Pahoehoe textured lava

Pahoehoe smooth lava   'a' a lava

Photo 15.1
Kilauea lava flow.

scope of the flow. Also, the photograph clarifies the appearance of the different types of lava, effectively complementing the text description.

Photographs now can be used readily to enhance almost any technical presentation. That was not always the case. For many years, the cost and difficulty of routinely including photographs in technical reports often discouraged or precluded their use. Today, however, modern digital cameras, scanners, and printing techniques have made the use of photographic images in technical reports easy and inexpensive. Image processing software also provides an effective means of easily enhancing, editing, and modifying images.

Here are two precautions for the use of photographs in technical documents:

• First, make sure each and every photograph serves a purpose relevant to the topic at hand. Do not just include a photograph because it is easy to do or because you need to pad the document (a favorite practice of some students).

- Second, when you are working with digital images, always ensure they have adequate pixel density for the purpose at hand. A density of 2 megapixels is adequate for normal use, while a density of at least 3 megapixels is preferred when you are enlarging images to fill a large part of a typical 8.5 × 11 page. (The author shot Photo 15.1 with a 2.1-megapixel digital camera while exploring the Hawaii Volcanoes National Park in 2002.)

Image Alteration

With modern desktop computer hardware and software, it is a simple task to combine, edit, and modify high-resolution images down to the pixel level. This capability can be used not only to enhance photographs, but also to "modify" reality—in effect, to create false composite images.

Photographic editing and composition is a powerful tool that can enhance the process of technical writing. Any edited image is usually, to some extent, a false image; however, that does not mean such images cannot and should not be used. On the contrary, they can be employed in the technical communication process as long as they are used properly. The appearance of objects can be altered to clarify a point or even show something that normally cannot be seen. (Photo 15.2 provides a cutaway X-ray view of the battery compartment in the author's laptop computer. The goal was to show the battery pack's location.)

Acknowledging Alteration

Digital imaging and the power of alteration can raise serious ethical issues. To avoid ethical problems, when the alteration of an image has fundamentally changed the information it represents, always acknowledge that the image has been

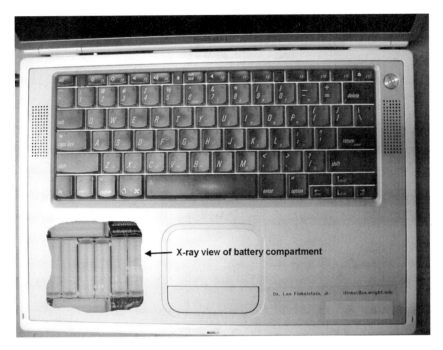

Photo 15.2
Composite image of computer.

altered in the caption or title for the image. Minor adjustments such as contrast, brightness, or hue are not normally acknowledged because they usually do not change the information that the image represents. For more on ethics, see Chapter 2.

Process Limitations

As mentioned earlier, when you are developing visuals, it is important to keep in mind the limitations of the intended output or production process for the document. These limitations are particularly important when you are using photographic images that will be printed. Since Photos 15.1 and 15.2 were intended for a single-ink production process, the original color information was discarded, and the photographs were converted to gray scale. This not only reduced the file sizes of the photographs, but also allowed a more accurate preview of how they would appear when printed. Remember, reproducing photographs requires that grayscale or color information be rendered properly in the final document, and this requirement alone often dictates special treatment and additional costs in the printing or duplication phase.

Conclusion

Visuals are a key element of modern technical writing. They aid immeasurably in describing complex topics precisely by providing a large amount of information.

Normally, we think of visuals as diagrams, graphs, schematics, tables, and images. Diagrams are drawings that show a mechanism or its components. Graphs, however, display and represent sets of numbers, values, quantities, and proportions. Schematics are visual representations of the structure of a mechanism or process, tables display the orderly arrangement of data or other information in columns and rows, and images or photographs provide accurate visual reproductions.

Checklist for Visuals

- Have I selected the best visual for the kind of information I am presenting?
- Have I accurately displayed the information?
- Have I numbered and titled each visual?
- Have I documented the source, if necessary, using a source line with the visual?
- Have I integrated the visual with the text discussion that references it?
- Have I referred to the visual prior to its placement in the report?

- Have I used labels in the visual that match the terms used in the text?
- If I altered a photograph, have I acknowledged that fact in the visual's caption or title?
- Have I considered the output publication process in my design of each visual?

CHAPTER SIXTEEN

Electronic Publishing

In the old days, publishing was fairly straightforward. The author would write something, the copy editor would polish it, the layout editor would design it, the typographer would typeset it, and the printer would print it. The output of the process was a pile of paper designed and printed in a particular way and then bound and distributed to those who were included on a list to receive it. The publisher controlled almost everything about the publication: the fonts, the layout, the general appearance—and, to some extent, who got it.

Information technology has changed all that with electronic publishing. The electronic distribution of printed material and the elaborate networking of information are part of the general move toward a "wired" society. Traditional printed materials (such as periodicals) are also being distributed increasingly in electronic form. In effect, paper and ink, as the primary communication medium, are being augmented, and to some extent displaced, by hypertext files and digitized images. Technical writing is part of this revolution in electronic publishing; technical writers today can no longer limit themselves to writing and preparing documents solely for traditional printing and duplication.

This chapter provides a brief orientation and introduction to electronic publishing, along with a few general considerations. The information and guidelines presented throughout this book are just as valid for electronic publishing as for traditional publishing. However, there are significant differences in the way one designs, organizes, and distributes material when it is being published electronically. These differences are the primary focus of this chapter.

What Is Electronic Publishing?

Electronic publishing is the process of distributing information, sometimes on a wide scale, through computerized storage media and networks. At its most basic level, it involves encoding and electronically distributing printed documents in file formats designed to be loaded into a computer for editing, viewing, distribution, and printing.

In many cases, such documents are distributed over the Internet. In other cases, these documents take the form of either e-mail messages or encoded attachments to e-mail messages, or may be carried in personal digital assistants (PDAs) as portable document or eBook files. These methods provide a fast, reliable means of moving complete files from one place to another; in fact, it is now possible to publish and distribute electronic documents more quickly than a traditional printing press can produce a single page. You can publish an electronic document by placing it on an Internet FTP or Web server, or by distributing it through an e-mail list server, so that potentially large numbers of people can access the file and use it.

Common Electronic Publishing File Formats

Document Files

Many different file formats are available for the distribution of electronic technical documents. Here are some of the more common ones, along with their file extensions:

- *Adobe Portable Document File* (.pdf) is widely used for forms, reports, and finished graphics. PDF files provide good quality and excellent cross-platform capabilities. In fact, PDF files are frequently the file format of choice for distributing documents or forms electronically because of their ability to maintain the exact formatting of the original material irrespective of the platform used.

- *Microsoft Word* (.doc) has become the de facto standard for word processing files. Word documents are supported by so many applications that they can be used effectively for cross-platform requirements; however, the formatting can vary from one computer and operating system to another, especially where special fonts, tables, lists, and graphics are included.

- *HyperText Markup Language* (.htm/.html) is used to distribute documents that can be viewed across multiple platforms and is the basis for Web publishing. While the appearance of the document can vary significantly from one platform to another, this file option does provide the functionality of hypertext links and virtually universal access for anyone with a Web browser. (See the discussion of hyperlinked documents in the next section.)

- *TeX* (.tex) is a specialized, complex typesetting system often used in professional technical writing where extensive mathematics are included. *LaTeX* is a macro that provides a system for managing the presentation of TeX documents.[1]

- *Rich Text Format* (.rtf) is designed to provide minimal formatting (e.g., bold, italic) for text in universal, cross-platform applications. RTF is limited in the formats and fonts supported and does not provide for integration of graphics with text. RTF can be particularly useful for transferring older word processing files whose formats are no longer supported by modern systems.

- *Plain text* (.txt) is text without any special formatting. Plain text is frequently used when the technical document file will be imported into a specialized system for formatting, or when cross-platform considerations preclude using other options.

Graphics Files

Graphics files used in technical writing vary widely and normally depend on the use and platform involved. Several file formats have been developed over the years to provide good cross-platform performance.[2] Here are some of the more common file formats for graphics.

- *Encapsulated PostScript* (.eps) files are used in professional publishing for graphics. EPS files are particularly useful when the publication process requires that images be scaled larger or smaller without loss of information. Encapsulated Post-Script files also contain a screen image for viewing and can be used to store virtually anything up to and including entire pages complete with fonts, layout, and graphics.
- *Tag Interchange Format* (.tif/.tiff) is a cross-platform file format for transferring bit-mapped images. TIF images are stored either as uncompressed files or as files with lossless Lempel-Ziv Welch (LZW) compression. Compared with other compressed-file formats, TIF images, originally developed by Aldus Corporation (now Adobe), can be relatively large but usually provide excellent quality.
- *Joint Photographic Experts Group* (.jpg/.jpeg) files are one of the most popular formats today for storing and exchanging graphics and photographs. JPG supports 24-bit color and provides substantial compression for continuous-tone, bit-mapped images. These files can also be used successfully for print publications, but only in their highest quality/lowest compression modes.
- *Graphics Interchange Format* (.gif) is a compressed-file format originally developed by Compuserve in 1987. The subsequent version (GIF 89a) is still popular for Web graphics and for applications in which file compression, multiple-image animation, and transparency are required. Since GIF does not support the color depth of .jpg files, it is often

used for line art or blocked graphics where relatively few colors are included.

- *Portable Network Graphics* (.png) is an evolving file format for use with Web browsers. PNG files are designed to be cross-platform. As the apparent successor to GIF, this format provides better color, compression, and image control than its predecessor.
- *Bitmap* (.bmp) is an uncompressed file format designed for use on Windows platforms only. Because its method of storing images is not standard and the files are relatively large, this format is not preferred for most technical writing applications.
- *Picture* (.pct/.pict) is a lossless-compression file format designed for use on Apple Macintosh platforms. It supports both bit-mapped and vector graphics and is normally employed for video editing, multimedia, and animation; however, .pct files are not used commonly in print publishing.

Hyperlinked Documents

A common form of electronic publishing encodes and electronically distributes information in a totally different fashion from what would normally constitute a printed document. Often, this type of publishing uses hypertext documents encoded with standardized, cross-platform markup languages. Hypertext is a method of representing words, images, sounds, and other online resources in a way that machines can read, and that allows people using these machines to move directly from one text component to another by pointing and clicking (usually with a mouse). Web browsers, such as Microsoft Internet Explorer and Netscape Navigator, interpret these hypertext languages. In doing so, they allow Web surfers, using hyperlinks, to move rapidly from site to site, page to page, and within a page, accessing text, graphics, tables, and diagrams as well as playing sounds and video.

Several markup languages exist. For example, the Standard Generalized Markup Language (SGML) is an international, professional standard for information processing and office systems. The more common HyperText Markup Language (HTML) is the standard language of the World Wide Web and is the language that all Web browsers use. Other markup languages also exist. Extensible Markup Language (XML) was developed from SGML and provides expanded capabilities for sharing data and supporting large-scale electronic publishing activities. Virtual Reality Markup Language (VRML) provides the capability to transmit and represent moving images of 3-D objects.

In addition to the Internet and other computer networks, hypertext documents can be distributed in many ways. Optical storage media are commonly used to distribute electronic files. For example, a compact disc (CD) can typically store about 700 megabytes of data, while a digital versatile disc (DVD) can store up to 17 gigabytes of data. The blue-laser DVDs now in development will provide many times that storage capacity. Given the capacities of these devices, it is a simple matter to publish an entire set of books on a single disk, including multimedia graphics, sounds, and full-motion video. Today's DVDs not only have the capacity to store all those books, but also can hold full-length, wide-screen motion pictures, with sound tracks in multiple languages. Next-generation DVDs will be able to do far more.

Many technical documents are now distributed in fully searchable, hypertext form on CD-ROM and DVD-ROM. For example, Microsoft provides professional-level technical support through a set of CDs or DVDs called *Technet*.[3] These disks contain an extensive database of instructions, technical notes, and seminars, as well as a current knowledge base of data developed from troubleshooting experience. The CDs and DVDs contain not only descriptions of problems and solutions, but also in many cases the actual software patches and drivers to implement those solutions.

In addition, user-level technical support is frequently provided by computer and software manufacturers in hypertext files embedded within application programs. Normally called *online help* or *balloon help,* these technical support files provide a complete online user's manual. Unlike printed manuals, online help manuals are readily available and can be searched by phrase, keyword, or full text. They also can be changed easily and inexpensively with every update of the software package.

Producing Hypertext Documents

Producing hypertext documents is not difficult, but it requires some understanding of computerized media and how these media work. Hypertext documents contain the same information as traditional documents but use a different philosophy of organization. Much of this difference relates to the nature of electronic media and how the user searches for and links to desired information.

Normally, hypertext documents are similar to those described in this book but are organized into parallel, interrelated structures. This approach is unlike the traditional way of doing things, in which documents are organized sequentially, reflecting the physical assembly of their pages.

Converting Traditional Documents to Hypertext

Many electronic writers initially create their technical reports in the traditional format, then convert them to hypertext by adding html tags. You will normally find it easier and quicker to use your word processor or Web page development programs to convert your standard document into hypertext. You may also want to divide the document and save it as several smaller word processing files first, then convert these files to html files that can be interlinked. The use of smaller file sizes reduces loading times, as will be discussed later. Or you can just use a single, larger html file and define internal hyperlinks within that document. In any case, you can use any Web authoring or editing tool to format and reorganize these converted materials and to establish the hyperlinks among or within them.

If you convert your technical document to html, you will find that the resulting document does not take advantage of the powerful linking capabilities of hypertext. You will have only documents that can be viewed as single pages by a Web browser. If you do not divide the original word processing document into subsections before converting it to html, you will wind up with one larger hypertext file that can be viewed only as a single page. Of course, you can still add all the necessary links whether you are working with a single file or multiple files.

Your new hypertext page probably will not look like the original document when you view it with a browser. Browsers handle formatting differently, and the formatting capabilities are much more limited than with traditional word processing and desktop publishing. To some extent, this really does not matter because you will probably need to rework your document anyway. Converting your word processing documents to hypertext files represents just the start of the electronic publishing process.

Next, you will need to set up hyperlinks in your document to provide parallel, relational access to each section. The goal is to create links that allow the reader to quickly and effortlessly move between parts of your document. The easier you make it to locate and go to key parts of your document (or link to other documents), the more effective your design will be. But do not go off the deep end creating hyperlinks! They can quickly add more complexity and confusion than they are worth.

How Web Browsers Work

When you are creating hypertext documents, it helps to understand how a Web browser displays information. When you tell your browser to display something, you normally do so by entering a Web address, called a *universal resource locator* (URL), or a local html file name and path on your hard drive. Your computer must load the file that is going to be displayed, along with its associated resources (graphic images and so on), into your computer's memory. Only then can the file be displayed completely. In other words, if resources such as photographs, diagrams, or charts are part of the file to be displayed, the browser must read all these resources into memory, as well as the html file itself, before the complete page can be viewed.

The requirement that browsers load all files is important when you organize your material. If your reader is accessing the information directly from disk or over a high-speed data network connection, this is no big problem. But what if your reader is accessing the file over a modem? Or what if the network is busy? By including excessive graphics in a single page, you unwittingly may make your page virtually unloadable for some users. Think about that when you are building large hypertext documents or including large graphics in your documents. Normally, it is much better to include links to several smaller documents than to put everything in a single file.

Another thing to keep in mind is that once you convert a technical document to hypertext, you lose some degree of control over its formatting.

The receiving computer's browser will control how your document appears. Even though some standardization exists with default font sizes and colors and the way html tags are handled, the displayed results still vary between browsers and video screens. That is why, when you are designing hypertext documents, it is a good idea to view them by using several different browsers, platforms, and operating systems to ensure that the results are acceptable. Some browsers allow users to configure their own formatting. If a reader of your document has weird tastes in fonts and colors, your document might wind up looking quite different than what you had in mind!

Guidelines for Organizing Hypertext Documents

Traditionally, technical documents are organized sequentially, from the first page to the last page, because that is how pages go together. As readers move through a technical report, they pass in an orderly way from one section to the next section, and within sections from one subsection to the next subsection.

This format works well with mystery and suspense novels. In technical reports, however, there is no mystery or suspense (at least, there shouldn't be!). So we try to mitigate the limitations of sequential access by adding devices such as indexes, dividers, and tabs—all of which are designed to give the reader direct, parallel access to sections or subsections. These devices allow the reader to more readily locate the desired information. In hypertext reports, we do not need to include such devices because hyperlinks and search capabilities can take readers quickly and precisely to the information they are seeking.

The trick to electronic publishing, then, is to properly organize the material to expose the maximum number of hyperlinks, at any given time and place, to the most essential information in the report. This technique is particularly important at the top level of an electronic report. All the important links should be visible, or obviously available, when the top-level page is displayed. In other words, you should see the links when the page is loaded, or you should have a clear indication that other links exist and that you can scroll to them.

To understand how to reorganize a traditional report into an electronic hypertext report, first go back to Chapter 9 and look at the 16XL1000000 laboratory report. The traditional document's organization is shown schematically in Figure 16.1. Notice how the traditional arrangement of the document is sequential, reflecting the successive way the pages are physically assembled. The flow starts at the beginning with the introductory statement of purpose, problem, and scope; and then it proceeds through the background discussions on theory and prior research. Test and evaluation methods are discussed, including the apparatus and procedure. This is followed by the findings (data) for power, distortion, and noise, along with an interpretation of the results. The conclusion provides an assessment and recommendation. We could mitigate this linearity with several techniques and devices, including a table of contents, an index, and dividers with tabs to help the reader locate the individual sections of the document. But the document would still be assembled sequentially.

Consider how we might reorganize this material when putting it into hypertext form. We are no longer constrained by the sequential assembly of physical pages, so we can think about putting important information in parallel form. Figure 16.2 provides a schematic of how this kind of parallel arrangement might work. Instead of sections of the report dictated by the physical requirements of page assembly, we now have screens that can be interconnected through hyperlinks.

Introduction

Purpose   *Problem*   *Scope*

Background

Theory   *Research*

Test and Evaluation

Apparatus   *Procedure*

Findings

Data   *Interpretation*

Conclusion

Assessment   *Recommendation*

Figure 16.1
Traditional organization.

Remember, the goal is to get all the major links on the top-level screen. Figure 16.2 shows, in a simplified way, the parallel hyperlink structure for the first two levels: the top-level screen and the screens immediately below this level. In reality,

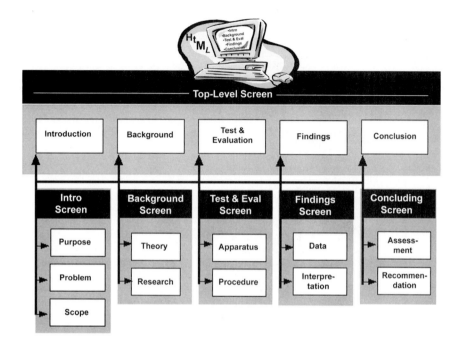

Figure 16.2
Hypertext organization.

this electronic document would contain several more screen levels, with many more hyperlinks pointing back and forth among all the significant sections—and perhaps directly to information embedded within sections and subsections, and to external Web resources.

The design of the page can be quite varied and, to a great extent, is limited only by your imagination, creativity, programming capability, and common sense. Modern Web authoring software, such as Macromedia Dreamweaver, typically comes with a variety of templates and color schemes. You can also surf the Internet to get ideas for layout and design. However, the best rule for electronic reports is to keep them simple.

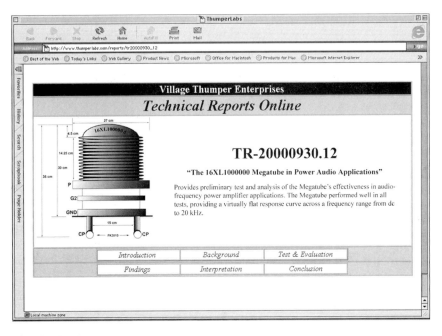

Figure 16.3
Top-level page.

Figure 16.3 provides a top-level screen for this electronic laboratory report on the 16XL1000000; the figure shows what the page might look like if it was viewed with a Web browser. The six buttons at the bottom of the page provide primary parallel links to the five major sections of the report and include one additional button for data interpretation, an area of primary interest. Each of those sections would have its own page, which, in turn, would contain the information or provide hyperlinks to additional pages, or to other websites containing additional information. Links could also be set up to go directly to specific pieces of information. Of course, all pages would contain links to other major pages as well as links back to the top-level page.

As discussed in Chapter 14, copyright laws provide federal and international protection for the expression of ideas, but not for the ideas themselves. Before electronic publishing and the Internet, the expression of ideas basically took the form of images printed on a page, whether those images were text, photographs, or illustrations. In other words, the ideas were encoded into the traditional media and were protected in that form. In many cases it was impossible to separate an idea from its expression (as with, say, a photograph)—and publishers exercised substantial control over what the expression looked like, what form it took, and where and how it was distributed.

About Electronic Publishing and Copyright

In effect, the limitations of the medium provided reasonably good protection for the ideas it carried. You could copy or duplicate a printed page, an illustration, or a photograph—but you had to have an original or good-quality copy to do so. After one or two generations of duplication, the copy would become so degraded that it would not be worth having. And, of course, copying usually is not free. These inherent limitations helped to prevent the widespread pirating of copyrighted material that we are seeing today.

Now, when an electronic document is placed on a Web server, it may well be accessible to anyone anywhere in the world. Viewing it requires that it be downloaded and copied to the viewing computer's memory—and probably, as well, to the browser's cache on the hard drive of that computer. Anyone viewing a Web document does not need to make a local copy because, given the nature of the technology and the way Web browsers work, that copy already exists. One needs only to specifically name and save it as a file to permanently retain an exact duplicate of the original. The html, graphics, and sound files are just as good, qualitywise, as those on the original server from which they were taken.

Consequently, copyrighted material on the Web is often available to anyone who wants it, and copies can spread rapidly with little or no degradation and virtually no control. So consider any technical report you place on a Web server to be public material in a global information infrastructure; and expect that people will look at, copy, and use it freely. Of course, you could encrypt the documents, password-protect them, or even place the materials on a secure server. But these inherent limitations to access may undermine the purpose of having the materials online in the first place. Moreover, no protection scheme is totally foolproof.

Other possibilities for protecting electronic documents exist, including digital watermarks and various licensing schemes. But when all is said and done, the Web is still a global community—and the rules and values you embrace may not apply everywhere. For copyright protections to be effective, they must be international, specifically applicable to online information, and enforced for all users who have access to the information. That may be where we are now headed, especially with the Digital Millennium Copyright Act (DMCA),[4] but we are not there yet. So think carefully about what you put online and why you are putting it there, before you actually do so.

Checklist for Electronic Publishing

- Have I properly divided the word processing file into subfiles before converting the document to html?
- Have I considered user capabilities and loading times when designing my document?
- Have I viewed my document, using more than one browser, platform, and operating system to check the formatting?

- Have I designed the top level to expose the maximum number of major links?
- Do I have a good reason for publishing this report electronically?
- Have I fully considered the copyright implications of putting my report on the Web?

1. For a discussion of TeX and LaTex, see Robin **Notes**
   Fairbairns, TeX FAQ, Internet: http://www.tex.ac.uk/
   cgi-bin/texfaq2html?label=whatTeX, May 13, 2003.
2. A good source for information on graphics file formats
   can be found at Internet: http://www.dcs.ed/ac.uk/
   home/mxr/gfx/.
3. For more on Microsoft's Technet, see the Technet
   home page at Internet: http://www.microsoft.com/
   technet.
4. The Digital Millennium Copyright Act (DMCA) addresses
   several issues raised by digital media, particularly
   computer networks, and especially the Internet.
   Among other things, the law protects system adminis-
   trators from liability for certain materials that may pass
   through their systems; better defines what constitutes
   copyright infringement on computer networks; length-
   ens the term of protection for copyright holders; and
   outlaws certain devices designed to circumvent copy
   and copyright protections.

CHAPTER SEVENTEEN

Presentations and Briefings

If you believe your public speaking days are over simply because you have completed your last required speech course, you have a rude awakening ahead. The competitive world of science, technology, and business is built around technical presentations. Increasingly, companies are relying on their technical experts to present information to various audiences. The topics can be quite varied, from plans and programs, to goods and services, to advanced theories and concepts. The audiences can range from your coworkers in a staff meeting to thousands of people in an international video teleconference.

What Are Presentations and Briefings?

Presentations and *briefings* are interpersonal performances in which concise technical information is provided to an attending audience. These performances, normally done live, are augmented by visuals and other media. Making effective technical presentations and briefings is not easy. Doing so requires a fertile imagination, a bit of courage, technical expertise in the subject, and the ability to communicate accurately and effectively.

When you are asked by your boss to make a technical presentation—and sooner or later you will be—do not despair. Wasting energy on fear will not help you do a good presentation. What will help is the following:

- Identify the purpose for the presentation, and locate or develop the kind of information required.
- Find a simple way to organize these ideas clearly and effectively.
- Gear the presentation and its materials to the audience and purpose at hand.
- Rehearse the presentation several times.

What makes a technical presentation effective is no big secret. Here are the key factors:

- Substantive ideas with precise, accurate information.
- Clear, coherent organization.
- Terminology and concepts appropriate for the audience at hand.
- Effective supporting media.
- Professional personal performance before the audience.

Substantive Ideas

The first consideration when you are giving a technical presentation is to have something worth presenting. In technical presentations the subject matter and purpose are usually well defined. You might be presenting a proposal on behalf of your company, the status of your project, the results of a laboratory study, or just information from a research project. In any case, the purpose and topic will dictate what substance needs to be included.

Clear, Coherent Organization

Once you figure out what you want to say, organize the material so you can present it effectively. Oral presentations differ from written ones in that spoken words are transient. If those reading your writing have trouble following what you are saying, they can reread it, think more about it, maybe ask someone else, and perhaps, in time, figure it out.

Those listening to an oral presentation, however, do not have that luxury. What happens if the members of your audience get lost or cannot follow your line of thought? What happens is mostly bad! They may stop to think about it, in which case they will not be listening to what you are saying next. Or they may ignore what you said and try to keep up, in which case they will have missed your earlier point. Or they may just get frustrated and stop listening altogether, in which case they are no longer part of your audience. Of course, audience members can always interrupt to say they do not understand; but in many situations, that is unlikely. It requires a tacit admission on their part, in front of their bosses and peers, that they do not understand what everyone else *seems* to understand. Normally, audience members keep quiet in situations like this.

That is why, especially for oral presentations, it is important to organize what you have to say in a way that is clear and obvious to your audience. Ideas—even difficult theoretical concepts—are much easier to follow and understand if they are organized logically and coherently.

Terminology and Concepts

Use words that are appropriate for your audience and type of presentation. Informal presentations can use informal language, whereas formal ones should be more "proper." Never use words that your audience will not understand. Readers can stop and look up something that they do not understand. But if your listeners do not accurately comprehend your words, you will lose their attention. Of course, they may also misunderstand what you are saying. This situation can be worse than when the audience does not understand, because now your listeners may believe they understand when, in fact, they do not.

Effective Delivery

You have figured out what to say, you have organized the material into a coherent and logical structure, and you have selected the proper language. Exactly how you deliver your presentation to the audience depends on the rhetorical situation. If you are speaking in an informal meeting, you might stand up and talk from your position at the table, or you might even stay seated. If you are making a formal presentation, you will probably do so from the focal point of the room, probably a podium or table up front. If you are giving a briefing, you will probably have charts—usually either transparencies with an overhead projector or computer-generated presentation graphics running through a video projection system or large display terminal.

You may find yourself speaking unassisted to a small group, or you may use a sound system with thousands of people in the audience. Perhaps you will have a time limit. You may flip your own charts, or you may remotely signal a technician in a projection booth to change your charts. And, increasingly, you may speak through a teleconferencing camera to many people in geographically remote locations.

Obviously, you must adapt your presentation to whatever situation you are in. However, some general principles apply to any presentation. First, make sure you look and act professional. Also, if you have briefing charts, make sure your charts are professional in content and appearance. Some listeners may interpret substandard charts or inappropriate personal appearance as a lack of interest or capability on your part.

One final note regarding delivery of your presentation: It is normal to be nervous. To some extent, this nervousness is a positive thing. When controlled, it can give you an edge that will make your presentation more lively. However, nervousness, when not controlled, can distract your audience and degrade your presentation. It can cause you to stumble over your words, lose track of time, perspire on your notes, and speak too rapidly.

So what do you do for nervousness? First, get to know your material backward and forward. Be familiar not only with what you plan to say, but also with the theory and details behind what you are saying. Second, rehearse your presentation—on your feet and out loud—to the point where you are comfortable giving it. Practice may not make perfect, but it helps build your confidence and capability.

In the technical world, you will generally see three distinctly different types of speaking situations. Here they are, along with a few hints for dealing effectively with them.

Speaking Situations

Impromptu

Picture this: You are a systems administrator working for a large company. You have been invited to attend the morning staff meeting. One of the vice presidents mentions hearing on the morning news something about a hacker breaking into a competitor's Web server and substituting a clown's picture for that of the CEO. Everyone laughs—except your company's CEO. With a deadly serious look on his face, he turns toward you and asks you to describe the security measures you are taking to counter such threats. Everyone turns, looks at you, and waits. You are on!

You have found yourself in an impromptu situation, where you have to talk intelligently on a complex topic with virtually no preparation time. Although such speaking would be highly unlikely in a formal situation, it does occur frequently in such informal settings as staff meetings. Clearly, this situation is risky. The wrong choice of words, topics, and arguments can be hazardous to your career. And if you say something stupid, people will not remember that you did so in an impromptu mode—only that you said something stupid. So here are a few suggestions for handling impromptu situations.

First, if you think there is a chance you will be put on the spot, think through what you will say in advance. Clearly, as a systems administrator, you should have known about the hacker; and once you were invited to the staff meeting, you should have realized that the topic might come up.

Second, when you are unexpectedly put on the spot, do what forensic coaches teach their competitive speakers to do: Buy some time to think about it. One way to do that is to divide the topic in some generic way. For example, everything has a past, present, and future. So, in this case, divide computer security into the past, present, and future—and talk, by way of introduction, about how it used to be in the days before networks and hackers. While buying time, you should be able to gather your thoughts for what you are going to say next. Of course, this assumes you know what you are talking about. If you are not sure, it is best not to make up facts in an attempt to bluff your way through the presentation. If you get caught, you will lose your credibility, which is difficult to get back.

Extemporaneous

Extemporaneous speaking is the preferred mode for a technical presentation because the presentation is well prepared but not precisely scripted. If you give an extemporaneous presentation, you will follow an outline, but you will use your own words to discuss the material.

Extemporaneous speaking can be effective in technical situations, but only if you know the information and have practiced the presentation. You can use note cards for facts and figures and outlines for the presentation, but do not read from full pages of text. It is easy to get lost in a full page of text; and when reading it, you will often come across as stilted and insincere.

Manuscript

Manuscript presentations are totally prepared in advance. When you give one of these, all you do is read the script, with maybe some gestures and inflection added at appropriate places in the manuscript for emphasis. Try to avoid these kinds of presentations. They come across as insincere and stuffy, and when lengthy, they often generate boredom and despair among those stuck in the audience.

However, manuscripts do have a place in technical presentations, especially where high precision in detail and word selection is essential. For example, technical presentations that are going to be translated into several languages need careful word selection, especially where international business and commerce are involved. Also, if you are making a legal statement on behalf of the company, it is usually best to read, verbatim, what the legal staff has prepared.

One more point: Never try to recite a presentation from memory. If you are distracted or have any kind of memory lapse, you will find yourself "hung out to dry" in front of an audience with nowhere to go and nothing to say.

Knowing the purpose for a technical document is critical to how you write it. In the same way, knowing the purpose for a technical presentation is critical to what information you include and how you present it. Generally speaking, technical presentations are informative, demonstrative, or persuasive.

Speaking Purposes

Informative

In an informative presentation, your primary goal is to give the audience facts and other information. Informative presentations often take the form of a background briefing, where no decisions are required and no particular response is expected from the audience. A briefing on the capabilities of the 16XL1000000 Megatube would be an informative presentation. As you might imagine, these are relatively nonthreatening events that normally are considered low-risk.

Demonstrative

The primary goal of a demonstrative presentation is to show the audience how to do something or how something works. Teaching a computer class how to access and use a simulation on the Web would be such a presentation. These kinds of presentations often require audience interaction. They also tend to depend on having tools, equipment, and materials available during the presentation. Consequently, while relatively nonthreatening to the audience, they can be higher in risk because of your dependence on the equipment's working properly when needed. Test the equipment thoroughly before the presentation, understand what you are doing, and have a backup plan should something go wrong. Also, make sure you take all safety precautions, if warranted, so that you are not endangering anyone by your demonstration.

Persuasive

A persuasive presentation tries to convince the audience to make a particular decision or take some specific action. Often these presentations take the form of decision briefings, where you ask the boss to fund your project or approve an organizational change. Of all the technical briefings, these are potentially the highest risk because scarce resources, such as dollars and people, are often involved. Your briefing may be part of a zero-sum game; if you win, someone in the audience loses. The best advice is to know what you are talking about, have supporting facts and figures readily available, and keep your cool under fire.

Technical Briefings

Technical briefings are focused oral presentations that use visual aids normally referred to as charts. These charts can take the form of slides, transparencies, or computer-generated graphics. The briefing charts provide an outline of the presentation and, like technical documents, include words, illustrations, photographs, line graphs, and tables. They also add visual interest and transitions to the presentation.

General Guidelines

Use briefing charts to punctuate the presentation with short phrases and visuals, but do not attempt to provide a manuscript on the screen for the audience to read. Here are a few tips for producing briefing charts:

- Take full advantage of computer-generated presentation graphics whenever possible. Programs such as Microsoft Powerpoint have become the standard for business and technical presentations. Such programs are powerful software packages that provide many sophisticated capabilities, yet they are also generally simple to learn. You will find that producing truly professional presentations, using standard templates and color schemes, is easy and inexpensive with these software packages. You can also use these programs to print high-quality paper copies (often called *hard copies*), as well as backup transparencies, just in case the computer system goes down when it is your turn to speak.

- Make sure your charts will be readable in the room in which you will be speaking. Normally you will want to use at least an 18-point font. Avoid script and fancy fonts because they can be difficult to read. Standard templates that come with software packages such as Powerpoint generally provide readable color schemes and font sizes.

- Pick your colors carefully. Lower-contrast combinations (such as light blue on darker blue) may look fine on your video screen and on quality projection systems, but could visually fall apart with lower-quality systems. This problem can also occur when the ambient light in the room reduces the effective contrast of the projected image. Additionally, avoid light-colored fonts on dark backgrounds if you are not sure of the projection system and room. Under good conditions they can look dramatic; however, you will find that dark fonts on light backgrounds are readable in marginal situations when inverted combinations are not. Keep in mind that stronger colors such as red can overpower some members of your audience, while weaker colors such as yellow may fade out.

Like technical documents, technical briefings are straightforward and easy to organize. Typically a technical briefing contains the following charts:

1. Title chart
2. Overview chart
3. Discussion charts
4. Summary chart
5. Concluding chart

Programs such as Powerpoint provide separate layout masters for title charts and the slide charts used in the body of the presentation. They also provide numerous color schemes and template designs, all of which can be customized with your own color preferences, artwork, and images. Pick a standard template, or develop your own specific color scheme and design. Then use that look consistently throughout the entire presentation.

Title Chart

The title chart leads off your briefing by telling your audience the topic of the presentation and your name, position, and affiliation. In some cases you might want to add your e-mail address and telephone number. Figure 17.1 provides a sample title chart done in Powerpoint. The charts in this figure and those that follow feature the logo of the Village Thumper Laboratory overlaid onto a standard Powerpoint design. Notice that the name of the briefer, Anita M. McFinkel (a fictitious cousin of the fictitious William E. McFinkel), has been included on this chart, along with her title and position.

Figure 17.1
Title chart.

Overview Chart

The overview chart lists the main topics to be discussed in your briefing. It describes to your audience how you are organizing your presentation and what information will be included in it. Be sure to use short phrases (called *bullets*) on this chart, not complete sentences. Figure 17.2 provides a sample overview chart.

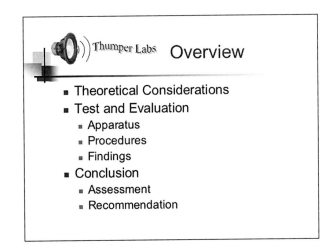

Figure 17.2
Overview chart.

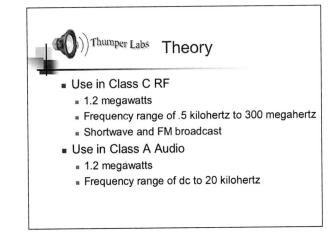

Figure 17.3
Discussion chart
(theory).

Discussion Chart

Discussion charts constitute the body of your presentation. Normally you will have one or more discussion charts for each topic listed on your overview chart. Figure 17.3 provides a sample discussion chart that uses short phrases (bullets) rather than complete sentences. Figure 17.4 provides a sample discussion chart that uses a line graph to present data.

Summary Chart

The summary chart gives your audience a brief summary of the important points of your presentation. In many cases you may be able to reuse much of your overview chart as your summary chart. Figure 17.5 provides an example of a summary chart.

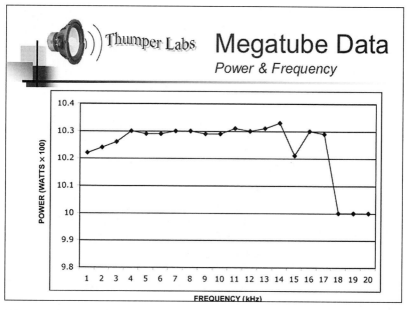

Figure 17.4
Discussion chart (graph).

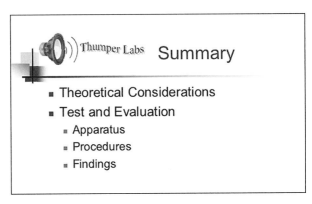

Figure 17.5
Summary chart.

Concluding Chart

A concluding chart may or may not be required. In presentations where you are concluding something from your material and perhaps even making a recommendation, you will need to add this chart. Figure 17.6 provides a sample concluding chart.

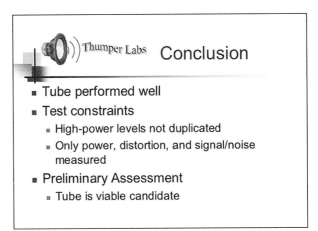

Figure 17.6
Concluding chart.

A guiding rule for producing technical reports is to use only the information that is necessary to get the job done. This is even more important in technical briefings. Avoid adding anything to your briefing that does not have a specific purpose or serve a necessary function.

Controlling Complexity

Visuals and Complexity

Your audience can absorb only so much information on the screen at one time. A service technician can effectively use, say, a large wiring diagram of a complex control system when that diagram is on his or her desk, and he or she has the time and need to trace through it. That same technician would find the same diagram useless, and even irritating, if you projected it onto a screen in a briefing. The detail would be difficult, if not impossible, to see or comprehend on the screen.

The same thing is true for complex tables or graphs of data on one chart. What if someone put up Figure 17.7 on the screen and then talked about task analysis for 30 minutes? The multiple line graphs are so visually complex, and the font size of the labels on the diagram is so small, that not even an eagle sitting in the room could read it, much less a human being. And the human being and eagle would have about the same level of interest and understanding. A good rule of thumb is that if you have to drop the font size below 18 points to fit the information on the chart, then you are probably putting too much information on the chart.

The best approach is to decide whether all that information is really necessary for the purpose of your briefing. If not, get rid of what is not needed. For example, in Figure 17.7, what does the specification diagram have to do with task analysis? Why is that diagram even included? Additionally, why do all the tasks need to be graphed on the

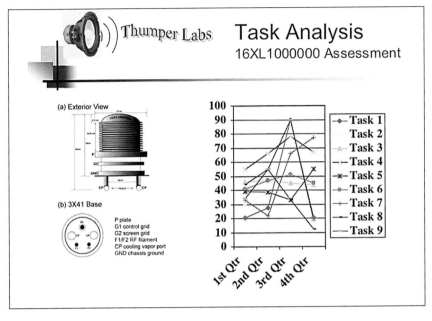

Figure 17.7
Overly complex chart.

same chart? Remember, more charts with less information are usually better than fewer charts with more.

Special Effects

Another way of adding complexity to briefings is by incorporating special effects. In presentation software packages such as Powerpoint, it is a simple matter to add animation, video, and sound to your charts. You can have text flying in and out, diagrams dissolving into one another, animated arrows directing the audience's attention, and a whole range of other visual effects. You can punctuate every visual effect with a pop, bang, or squawk. You can stir the emotions with big band music and soothe them with the sounds of the ocean.

The question here is not so much what you can do, but why you would want to use these effects. Special effects can be useful in small quantities—but be careful! Almost all of these effects get old quickly, even for the most patient audience. For example, using flying text or a humorous sound effect to highlight a point is fine, but using such effects throughout the entire briefing will become irritating and distracting. Too much of anything is usually bad!

There is one final consideration for special effects. Many of these effects are data-intensive, requiring a great deal of processing overhead by the computer. That may not be a problem when you are sitting at home or in the office designing your briefing on your top-of-the-line computer. But what happens when you take your presentation on the road and the only available machine is old? Your presentation may be unusable. The same thing can occur when you use high-resolution graphics in your presentation: you may have to wait for what will seem an eternity while the machine loads the next chart.

General Checklist

- Do I understand the occasion and purpose of the presentation?
- Am I presenting information that is substantive and relevant?
- Have I organized my information effectively?
- Am I using language and media that are appropriate to my audience?

Checklist for Technical Briefings

- Have I included a title chart with my name and organization?

Checklist for Presentations

- Have I included an overview chart listing the main topics of my briefing?
- Have I included at least one discussion chart for each topic on my overview?
- Have I included a summary chart (and conclusion chart if necessary)?
- Have I ensured that all my charts are readable when projected?
- Have I chosen an effective design scheme, and am I using it consistently?
- Have I avoided unnecessary complexity in any of my charts?
- Have I evaluated the briefing room and equipment beforehand (if possible)?

18

Resumes and Interviews

The best way to get a job is to get your foot in the door by knowing someone important who is already inside. This approach is frequently called *networking.* When it is time to go out and get a job, or to move from one job to another, networking with the right people in the right places can be extremely effective. Unfortunately, for those of us who may not know these people, that approach will not work. If you fall into the latter category, you will have to start by developing a resume and cover letter. With these documents you can earn the opportunity to interview for a position and land the job.[1]

Generally speaking, resumes are not all that useful because they just do not work well. Few people who send out resumes actually get a job as a result. But that is generally speaking. For engineers and scientists, resumes can be effective—at least in getting an opportunity to interview. In particular, engineering and computer-related science positions tend to be based on requirements that demand specific skills such as fluency in Unix, background in homeostatic control systems, hands-on experience with electron microscopy, and the like. These are objective, verifiable qualities that lend themselves to resume presentations.

Effective resumes must contain specific kinds of information that relate to the position desired. They are well written and are accompanied by an equally well-written cover letter. The resume and cover letter must be flawlessly edited. This chapter will show you how to produce an effective, professional resume and cover letter geared specifically to engineering and science positions. It will also give you some useful tips for what to expect in an interview.

What Is a Resume?

Resumes are specialized proposals by which you offer your services to fill a position. Normally resumes include a short summary of experience and qualifications; however, engineering and science resumes need to do more than that. A resume needs to match your specific skills clearly to an employer's needs.

Your resume should reduce an employer's uncertainty about the perceived risks involved in granting you an interview, making you an offer, and hiring you. Employers do not want to make the wrong decision and hire the wrong person. Those are expensive mistakes that employers try hard to avoid. Your resume should reassure them that they are not making a mistake with you.

Suppose a firm is looking for an entry-level electrical engineer to design microwave communication systems. You have the qualifications to do that job, but what kinds of information should you highlight to reduce an employer's uncertainty about hiring you? You would probably want to show the employer that

- You want to work for that particular employer.
- You want a position involving microwave communication systems.
- You have the needed technical background to do the job.
- You have the required written and oral communication skills.

- You have the necessary personal traits.
- You are available when the employer needs you.

The kind of resume we are talking about here contains more than just a short summary of your skills. You need to think of your resume as a sales document by which you convince a prospective employer that you are low-risk—that you want to work in the position and that you will do a great job. To accomplish this, your resume should have the kinds of information called for in Outline 18.1.

Writing a Resume

Writing a resume is demanding work—but because you are dealing with your career, you should be motivated to put forth the effort. Fortunately, you already know a lot about the subject; but determining your strengths and developing the document are still challenging. It is difficult to be objective when you are the subject. So take your time, and ask people you trust (but not necessarily close friends) to read and respond to what you are writing.

Outline 18.1 provides a simple, straightforward approach for organizing a resume.

Outline 18.1 Resume

Name	List your name, address, telephone, and e-mail address.
Objective	Describe the kind of position you want.
Strengths	Highlight your strongest skills and attributes.
Education	Document your formal education.

- Degrees
- Certifications
- Honors
- Relevant course areas

Computer skills	Document your computer literacy.
	• Languages (in descending order of fluency)
	• Operating systems (in descending order of expertise)
	• Applications (in descending order of expertise)
	• Platforms (in descending order of expertise)
Experience	(in order of relevance to job objective)
Personal	Provide additional, relevant personal information.

As Outline 18.1 shows, you should start with your name, address, and telephone number. In fact, this information should provide the heading of your resume. As an example, we will use William E. McFinkel, a person with solid qualifications. (Remember Sabrina McFinkel and Anita McFinkel in previous chapters—they are fictitious relatives of the fictitious William McFinkel.) Anyway, as part of your heading, it is a good idea to include an e-mail address if you have one, but only if you check your e-mail daily. If you do not check it that frequently, omit it since including the address might make you seem unresponsive to an employer trying to e-mail you. Because McFinkel rarely checks his e-mail, his e-mail address is not included.

William E. McFinkel
4141 E. 34th Street
Selingford, Ohio 45121
Home (937) 999-9999
Work (937) 999-0000

Objective

Next, McFinkel needs a job objective that should clearly relate to the position for which he is applying. The objective should also be specific enough to reassure the prospective employer that McFinkel really knows what he wants. Consider this job objective:

Objective
Seeking a tough, challenging engineering position in radio-frequency communication systems where I can help achieve corporate goals in a team environment.

What does this objective tell us about McFinkel? First, he knows what he wants. Employers like that, along with the fact that his objective matches the available position. Hiring someone is risky for any company; making a mistake can be costly and traumatic for all concerned. Because he knows what he wants to do, and because that seems to fit with the available position, McFinkel represents a good match for the employer. Consequently, it is likely that he will be happy working for the company, and that alone reduces the company's risk in hiring him.

Second, he is looking for a tough, challenging position. To an employer, he sounds like a real "doer"—exactly the kind of worker who will thrive on solving difficult problems for the company.

Third, he wants to help achieve corporate goals in a team environment. So he understands the nature of employment. He understands that companies exist primarily to make money. He also understands that he will be working in a team environment, which is important because technical problems are rarely solved by one person.

Strengths

McFinkel needs to list the qualities that make him special and set him apart from others. Notice that the following strengths statement is written in phrases, not complete sentences. Phrases are acceptable here because they are economical, to the point, and easy to read.

Strengths
Special blend of technical, interpersonal, and communication skills and experience. Hardworking,

creative, articulate. Able to work well with others to get the job done.

To an employer, this short statement effectively shows technical competence, the ability to work well with others, and effective communications as important qualities that distinguish this applicant.

Education

Frequently, technical positions have specific degree requirements. This requirement may be a function of company policy or may be mandated by a client with whom the company is doing business. For engineering and science positions, always highlight your degree, not your school; with rare exceptions, employers are hiring your education, not your college or university.

Here is McFinkel's educational information:

Education

B.S. in Electrical Engineering with University Honors (June 2005)
Wright State University, Dayton, Ohio 45435

A.S. in Engineering Technology (June 2000)
Ocean Breezes Technical Institute, Ocean Breezes, North Carolina 27334

Relevant Course Areas

- Microwave Systems
- RF Systems
- Digital Computer Hardware
- Computer Networking Control Systems
- Linear Systems
- VLSI Design
- Pulse and Digital Circuits
- Software Engineering
- Data Structures
- Computer Graphics
- Comparative Languages
- Technical Writing and Speaking for Engineers

Notice that the degrees are shown first, followed by the schools, locations, and dates. Also notice that in the list of relevant coursework, the course areas most relevant to the objective are given first.

Computer Skills

Computers are essential tools in almost any workplace, and particularly in the technical and scientific workplace. Many engineering and science jobs are provided by small companies, which usually place a premium on multidimensional employees. So it is useful to note any computer skills you might have, even though you are not necessarily applying for a computer position. Always looking for any competitive edge, McFinkel has done just that:

Computer Skills

Languages (descending order of fluency):
- C, C++, Ada, HTML, Pascal, Visual BASIC, COBOL, FORTRAN

Operating systems (descending order of expertise):
- Unix, Solaris, Windows 2000/XP, MacOS9/OSX

Hardware (descending order of expertise):
- Sparc, Intel clones and Pentium IV, and Macintosh G-3/G-4

Applications (descending order of expertise):
- Word, Illustrator, Photoshop, Excel, Powerpoint, Access, Navigator, Internet Explorer

Notice that McFinkel has some impressive computer credentials. (If you are going to invent a person, why not do it right?) However, understand that McFinkel is simply an example. Never invent qualifications on a resume! Lots of people lie on resumes, which is one of the reasons resumes may not be effective; however, where technical qualifications are concerned, what you claim is easily verifiable. For example, if you say on your resume that you are fluent in C or Ada, all an employer has to do to verify your claim is ask you to implement a routine in C or Ada. That kind of thing happens in many interviews, either formally or informally. If it happens to you and you cannot do what your resume says you can do, then the interview is over, and so is the job opportunity.

Experience

Nothing is quite as persuasive as a track record of success. Virtually any kind of successful prior employment can be valuable on a resume, but successful employment in a job that relates to the objective can provide a tremendous advantage. This is especially true if you can show that you exceeded the expectations of the job.

You do not need to list all your experience on a resume. On a job application form you probably will; but on your resume, just list the experience that relates directly to the job objective. Space permitting, you can also list representative experience that does not relate directly to the objective.

McFinkel's resume lists his technical experience that relates to the job objective. However, his resume does not include his 14 jobs as a "broiler boy" with various fast food restaurants, or his work for the city cleaning up around the stadium after baseball games. That experience is simply not relevant.

Experience

4/98–Present: Engineering Technician, Computing Imaging Corporation, Dayton, Ohio. Responsible for installing and maintaining Unix, Windows, and Macintosh computer systems.

- 2001 and 2002 Outstanding Employee of the Year
- 2002 and 2003 recipient of the IEEE Merit Scholarship

1/96–3/98: Repair Technician, Computer Service Center, Dayton, Ohio. Responsible for PC computer system software and hardware installation.

- 1998 Excellence in Maintenance Award

Notice that McFinkel's experience lists inclusive dates, job title, company, and location. It also summarizes the duties of the job and documents performance that exceeded expectations.

Personal

The final section of the resume is the catch-all, personal category. You need to include job-related information here that may enhance your value to the company. For example, if you are not a U.S. citizen, that is okay—but if you are, by all means mention it. That means the company can bid you on certain contracts with the government or defense-related firms. Also describe your availability—the date you will be able to start work. It is sometimes useful to indicate that you are willing to travel or relocate to meet company needs (if, of course, you are).

The final section of McFinkel's resume might look something like this:

Personal
U.S. citizen
Available June 2005
Willing to travel and relocate

---

1. *Be sure your resume is perfect both grammatically and stylistically.* Pay excruciating attention to detail: There should be absolutely no errors. Employers look for communication skills, and, fair or not, they will key on any grammar or style error as evidence that you do not write well. Take advantage of grammar- and spell-checking software, but do not rely on it totally.

2. *Tailor your resume for the position.* With word processing software, this is a snap. Remember, even good resumes often do not work, and generic resumes never work. Be sure to create

Ten Tips for Creating a Good Resume

separate resumes for different types of jobs. A resume created by a computer scientist for a software engineering position will not work well for a network administrator position. Specifically, gear your objective to the job, and present the most relevant and supportive data first.

3. *Always be truthful in your resume, but not self-effacing.* In other words, do not say, "I have a serious heart condition, but so far I have been stable with medication." Accentuate the positive about yourself, but do not go off the deep end. Do not write, "I am absolutely the most brilliant engineer, and I represent the quintessence of intellectual power and personal achievement." Use your common sense!

4. *Limit your resume to two pages.* If you can keep it to one page, that is even better. One way to keep it short is to avoid minor details that the employer already knows or will not care about. For example, "references available on request" can be assumed. Also, non-job-related data, such as the names or occupations of your family, or that you like to swim, dance, and work for world peace, are facts that no employer really cares about.

5. *Do not include religion, politics, or fraternal organizations.* This information can be risky and usually provides nothing relevant to the job objective. Additionally, it is usually a good idea to avoid mentioning hobbies, unless, of course, they are relevant to the job objective. For example, amateur radio might be a hobby that is relevant to a microwave engineering job.

6. *Do not use fancy paper or exotic folds, and do not paste a photograph of yourself on the first page.* Employers of engineers and scientists are usually very pragmatic and substance-oriented.

7. *Avoid acronyms generally.* Use acronyms only when you are absolutely sure your reader knows them.

8. *Never refer to yourself in the third person.* It comes across as stilted and sounds weird.
9. *Never put a date on a resume.* If you do, your resume will seem outdated almost immediately.
10. *Do not refer to salary requirements.* Salary is a topic better suited to the job interview, which normally includes filling out a formal job application form before the actual interview begins. If you are pressed to include something on salary requirements, then just say that your salary is negotiable, or express your salary requirements in a range.

Cover Letters

Whenever possible, send your resume with a cover letter. However, simply including a note that says, "Here is my resume; give me a job" will not be productive. In fact, this wastes one of the most powerful tools you have to reduce an employer's perceived risk of hiring you.

A good cover letter is every bit as persuasive as a good resume, and in some cases it can even substitute for one. A bad cover letter often winds up in the trash, along with the resume attached to it.

An effective cover letter does four things:

1. It demonstrates that you know about the company and really want to work for it.
2. It summarizes your key skills and experience that make you a desirable candidate.
3. It describes the personal traits that make you a desirable candidate.
4. It provides a conclusion that creates goodwill and invites a favorable response.

Look at McFinkel's cover letter, and notice how each paragraph has a particular function. First, start the letter with the name, address, and contact information used on the resume:

William E. McFinkel
4141 E. 34th Street
Selingford, Ohio 45121
Home: (937) 999-9999
Work: (937) 999-0000

Next, whenever possible, address the letter to a person, not an organization—and *not* "To Whom It May Concern." Sometimes you will not have a choice; but if you can find the name of the person running the office that has the job for which you are applying, use it. McFinkel's letter is addressed to the senior engineer who is supervising microwave systems development:

Dr. Tracy Ann Burger
Senior Engineer, Microwave Systems
E-Wave Systems International
595 Miner Road
Wutherford Heights, Ohio 44132

In the body of the letter, the first paragraph should demonstrate that you know about the company and really want to work for it. McFinkel has done some homework and shows that he is not just "shotgunning" this letter everywhere:

While doing career research to locate companies that may have electrical and communications engineering opportunities, I read about E-Wave Systems International's government work with advanced satellite communications systems. In fact, the more I read about your company and its activities, the more convinced I became that I'd like to be part of the E-Wave team. To that end, I have attached my resume with the hope that you will give me the opportunity to discuss further how I might fit in and make a contribution at E-Wave Systems.

Notice how this first paragraph both demonstrates McFinkel's knowledge of the prospective employer and, as a bonus, shows that he wants to be a team player. He has taken the time to learn about the company's work with the government on advanced satellite communications systems, and he clearly wants to explore how he might fit in with the E-Wave Systems team.

The next paragraph briefly summarizes McFinkel's skills and experience that make him a desirable candidate. Notice that he does not go overboard here because the details are available in the resume. This letter's goal is to get the employer to look at the resume.

> I am currently a senior in Electrical Engineering at Wright State University. My degree program, which I will complete with University Honors this spring, comprehensively blends electrical and radio-frequency engineering concepts and principles. In addition, during the past few years, I have worked as an engineering technician for Microwave Products Corporation, with primary responsibility for maintaining wave guide calibration systems. I believe this has given me valuable experience with the same technology being used by your company in its microwave design activities.

The third paragraph reviews McFinkel's personal traits that make him a particularly desirable candidate:

> Personally, I am a strong, internally motivated self-starter with excellent analytical and organizational skills. I pride myself on being able to work well with others in a team environment and on having proven abilities in written and oral discourse. My technical communication skills are particularly strong.

A point worth noting: This letter is well written, and that alone supports McFinkel's claim that his communication skills are particularly strong. If he came across as inarticulate, that claim would be compromised, as would the entire resume package.

Finally, the last paragraph provides a conclusion that creates goodwill and invites a favorable response:

> Thank you for taking the time to review my resume. I realize how many letters and resumes you must receive, but I know I can be a valuable resource for your organization, and I would welcome the opportunity to interview at your convenience.

Notice that McFinkel didn't say, "I want a job—schedule me immediately for an interview!" You would be surprised how many people actually write something this brash in their cover letters and then wonder why they never get a response.

Finding Jobs on the Internet

Increasingly, technical people are searching for jobs and posting their resumes electronically on various Internet job sites.[2] Some of these sites simply provide a public posting forum, whereas others actively match job skills to positions. Many sites are free to the job applicant, but others charge a fee—sometimes for each posting or search.

The efficacy of this online method varies considerably depending on job skills, regional demand, the particular site, and luck. Generally speaking, the information you provide online is similar to the information in a traditional resume; however, the format and specific content may vary. The best advice is to approach online job searches as a resource for finding employment but not as a replacement for more traditional job search methods.

Putting It All Together

Here is the complete resume package, including the cover letter and the formatted resume. Letter and resume formatting can vary, but it is generally best to use a traditional block or indented style that is normally available on word processors.

William E. McFinkel
4141 E. 34th Street
Selingford, Ohio 45121
Home: (937) 999-9999
Work: (937) 999-0000

Dr. Tracy Ann Burger
Senior Engineer, Microwave Systems
E-Wave Systems International
595 Miner Road
Wutherford Heights, Ohio 44132   March 23, 2005

Dear Dr. Burger:

While doing career research to locate companies that may have electrical and communications engineering opportunities, I read about E-Wave Systems International's government work with advanced satellite communications systems. In fact, the more I read about your company and its activities, the more convinced I became that I'd like to be part of the E-Wave team. To that end, I have attached my resume with the hope that you will give me the opportunity to discuss further how I might fit in and make a contribution at E-Wave Systems.

I am currently a senior in Electrical Engineering at Wright State University. My degree program, which I will complete with University Honors this spring, comprehensively blends electrical and radio frequency engineering concepts and principles. In addition, during the past few years, I have worked as an engineering technician for Microwave Products Corporation, with primary responsibility for maintaining wave guide calibration systems. I believe this has given me valuable experience with the same technology being used by your company in its microwave design activities.

Personally, I am a strong, internally motivated self-starter with excellent analytical and organizational skills. I pride myself on being able to work well with others in a team environment and on having proven abilities in written and oral discourse. My technical communication skills are particularly strong.

Thank you for taking the time to review my resume. I realize how many letters and resumes you must receive, but I know I can be a valuable resource for your organization, and I would welcome the opportunity to interview at your convenience.

Sincerely,

William E. McFinkel
Enclosure: Resume

William E. McFinkel
4141 E. 34th Street
Selingford, Ohio 45121
Home: (937) 999-9999
Work: (937) 999-0000

Objective
Seeking a tough, challenging engineering position in radio-frequency communication systems where I can help achieve corporate goals in a team environment.

Strengths
Special blend of technical, interpersonal, and communication skills and experience. Hardworking, creative, articulate. Able to work well with others to get the job done.

Education
B.S. in Electrical Engineering with University Honors (June 2005)
Wright State University, Dayton, Ohio 45435

A.S. in Engineering Technology (June 2000)
Ocean Breezes Technical Institute, Ocean Breezes, North Carolina 27334

Relevant Course Areas

Microwave Systems	Pulse and Digital Circuits
RF Systems	Software Engineering
Digital Computer Hardware	Data Structures
Computer Networking Control Systems	Computer Graphics
Linear Systems	Technical Writing and Speaking
VLSI Design	for Engineers

Computer Skills
Languages (descending order of fluency):
• C, C++, Ada, HTML, Pascal, Visual BASIC, COBOL, FORTRAN
Operating systems (descending order of fluency):
• Unix, Solaris, Windows 2000/XP, MacOS9/OSX
Hardware (descending order of expertise):
• Sparc, Intel clones and Pentium IV; Macintosh G-3/G-4
Applications (descending order of expertise):
• Word, Illustrator, Photoshop, Excel, Powerpoint, Access, Navigator, Internet Explorer

Experience
4/98–Present: Engineering Technician, Computing Imaging Corporation, Dayton, Ohio. Responsible for installing and maintaining Alpha-based, digital computer systems.
• 2001 and 2002 Outstanding Employee of the Year
• 2002 and 2003 recipient of the IEEE Merit Scholarship
1/96–3/98: Repair Technician, Computer Service Center, Dayton, Ohio. Responsible for PC system software and hardware installation.
• 1998 Excellence in Maintenance Award

Personal
U.S. citizen
Available June 2005
Willing to travel and relocate

The purpose of a cover letter is to get an employer to read your resume. The purpose of a resume is to get an employer to interview you. The purpose of an interview is to get an employer to offer you a job. If you have the technical skills and personal characteristics an employer needs, and if you did a good job with your cover letter and resume, you will probably be invited to interview for the position.

It is important to realize that interviews are stressful and risky for both the interviewer and interviewee. Both parties have a lot to gain or lose depending on how things go. Interviews are decision points: The employer has to decide whether to make you an offer, and you have to decide whether to accept it.

Interviews come in all shapes and sizes, and they may have different purposes. Some are screening interviews designed to narrow the field to two or three top candidates; these are often conducted over the phone. They are usually followed by in-person hiring interviews. The final decision to hire is normally made based on the results of this interview. If you make the cut in the screening interview, you will be invited to the hiring interview.

Interviews can be as short as 15 to 30 minutes or can last several days. Some require only that you answer questions, whereas others require you to make a formal presentation about yourself and your work. Interviews can be formal, where you sit down in a room full of important-looking people and respond to one or more interviewers' questions. Or they can be very informal, such as a discussion at lunch or dinner.

Two things should occur in any interview: The employer should size you up to determine if you can fill the position, and you should evaluate the employer to see if you want to work for that company. Note that when you report for an interview, you will usually be given a job application form to

Interviewing

complete. Be sure to have complete contact information for each of your references with you, along with your employment history and extra copies of your resume.

Here are some tips for successful interviewing:

1. Fully research the prospective employer before your interview. You should have done some of this research when you wrote your resume and cover letter. Now, however, you need more detailed information about the company, the kinds of work it does, and its track record of success. Also try to assess the quality of the people working there, how stable their employment is, what sort of salary and benefits package the employer offers, the cost of living in that area, and so on. For many employers, much of this information is readily available on the Internet or in a library. You can also request brochures and an annual report from the company before you interview. In some cases, you can make discreet inquiries with those you trust who have experience with the company or who know people who work there.

2. Be prepared to answer not only technical questions but also personal ones. For example, if you were interviewing for a civil engineering position involving the design and construction of highways, an interviewer might ask you about cyclical loading of concrete and asphalt, or fatigue and stress problems associated with aging bridges. But what if the interviewer asked you questions like these?

    -   "What are your greatest strength and greatest weakness?"
    -   "What are your long-term career goals?"
    -   "Where do you see yourself in 5 years?"
    -   "Why do you want to work for this company?"

- "How well do you work with others?"
- "What do you know about this position?"

How would you respond to these questions? Think about that—these are probably the kinds of questions you will be asked first. Many new engineering or science graduates have no problem with the technical questions; but when asked the personal ones, they either turn nervous and incoherent or say something unwise.

3. Be prepared to discuss your salary requirements. You can find a variety of salary surveys on the Internet and in libraries. Identify the going rate for your particular skill level in the region where your employment will occur, and decide, in advance, what your bottom line will be. On the job application form, you may be asked for your salary requirements. To keep from revealing your bottom line too early, just indicate on the form that your salary is negotiable or express your requirements in a range.

4. Listen carefully in the interview. Take your time and respond to what is being asked. If you need to think about the question for a moment or two, say so. Do not blurt out a careless answer in an effort to respond quickly to a difficult question. Also, if you are not sure what is being asked, request clarification.

5. Look and act professional at all times. That does not mean you have to rent formal attire for the interview, but you should dress properly. For some companies, that may require a coat and tie, but find out in advance and do what is expected. Also act professional in your demeanor and in the substance of your discussions. Be on time for your interview, do not make mindless comments, and do not speak disparagingly about former bosses and coworkers. Always be positive.

Cover Letter Checklist	• Have I demonstrated that I have researched the prospective employer and have a reasonable knowledge of the company and the position?
	• Have I indicated that I really want to work for this company?
	• Have I summarized my skills clearly and succinctly?
	• Have I summarized my personal traits that enhance my qualifications for this position?
	• Have I demonstrated my desire to discuss further how I might contribute to achieving corporate goals?
	• Have I indicated my willingness to interview for the position?

Resume Checklist	• Have I put my name, mailing address, telephone number, and e-mail address (if appropriate) at the top of my resume?
	• Have I provided a clearly stated job objective that is not generic?
	• Have I briefly but effectively described my strengths?
	• Have I listed my higher education, including degrees, honors, and coursework?
	* Have I summarized my computer skills?
	• Have I listed my experience, putting first the most relevant to the job objective?
	• Have I included a personal section that provides, as appropriate, information about my citizenship, availability, and willingness to travel and relocate?

Interview Checklist	• Have I researched the company, and do I understand the position?
	• Have I prepared myself to answer technical questions about what the company does?

- Have I prepared myself to answer personal questions about my career goals and how working for this company in this position fits in with those goals?
- Have I consulted salary surveys, and do I know my minimum salary requirements?
- Have I dressed properly, and do I have an appropriate appearance for the interview?

---

1. For an extensive discussion of resumes, cover letters, and interviewing, see Richard Nelson Bolles, *What Color Is Your Parachute?* Berkeley, Calif.: Ten Speed Press, 2003.
2. Internet career sites represent a dynamic resource that changes almost daily. However, some of the major career sites that have been around awhile include the following:
   - Internet: http://www.careerbuilder.com
   - Internet: http://www.careermart.com
   - Internet: http://www.careersite.com
   - Internet: http://www.collegegrad.com
   - Internet: http://www.monster.com
   - Internet: http://www.4work.com

Notes

19
10011

Team Writing

You find yourself in an engineering design course where you are grouped with three other students, none of whom you know. Your group is tasked to design a highly efficient, inexpensive, environmentally friendly snow and ice removal system to keep the engineering dean's parking place open during the winter months. Your grade in the course will be based primarily on the quality of your group's solution as documented in a technical project report. The report must be produced by your group and delivered on the last day of class. Welcome to team writing!

Team writing, often referred to as *group writing,* is the process in which two or more authors work together to produce a document or set of documents to fulfill a requirement. Team writing is basically a "good news–bad news" thing. The good news is that you have several people with varied backgrounds, skills, and abilities from which to draw. The bad news is that you have several people with varied backgrounds, skills, and abilities from which to draw. The extent to which the news is good or bad depends in large measure on whether you are part of a team or a group, because *group writing* is really misleading.

If you want to do well in the course, you need to be part of a team, not a group. A *group* implies only the existence of an aggregation of people. A *team,* however, is an aggregation of people organized and managed to achieve some purpose. This distinction may seem academic to some, but real differences exist between groups and teams—and understanding and dealing with these differences are essential if any group, including your group, is to prosper in the team writing environment.

The collective experiences, skills, and beliefs of the individual members can be a powerful force when properly managed, or working together can be characterized by hurt feelings, wasted resources, and, worst of all, poor quality. Turning the collective diversity and talent of a group of writers into a successful team of writers is not easy. It requires the best effort of all involved, the right blend of skills, a great deal of common sense, and effective management and leadership.

Student versus Professional Team Writing
Team writing may represent a real adjustment for anyone whose writing experience has been limited primarily to individual work. As a student, you have probably gained most of your writing experience while working alone in classes. Some courses, such as the group engineering design scenario mentioned earlier, do offer some team writing experience—and getting that experience is great! However, be aware that these courses provide a modified form of team writing. If you are one of several students selected at random and assigned a project by your teacher, typically you will approach the problem by getting together and determining who on the team is the best writer. In the end, that person may well become the "team writer."

The classroom experience differs from the real world of business and industry, where the complexity of work often requires that technical writing be done by many members of a team. The scope and magnitude of the material, along with the required expertise in multiple disciplines, frequently prevent any one individual from writing the entire document. In the case of formal proposals, for example, proposal managers may even head up several writing teams. These managers often oversee various experts in applicable disciplines who are working together to produce an effective proposal. In this case, technical experts who form a "technical team" may write the technical section, while functional or management experts may write the management section, and budget experts may write the cost section. In the end, these and other sections are pulled together to form one, coherent document—hopefully a winning proposal.

The Process of Team Writing

The team writing process consists of several steps that can vary significantly from one job to another depending on the requirements of the particular situation. For example, some teams are formed before the team leader is named, other teams elect their own team leader after being formed, and still other teams have the team leader designated by higher management. Some teams select writers for each specialty from within their group; others have professional technical writers assigned to them; and sometimes a single publications office in a company provides professional layout, design, and copyediting. So clearly, many variations may exist for any given situation. That being said, a generic model for the team writing process is provided in Outline 19.1. It serves as a general guideline for doing team writing and includes three steps: determining requirements, taking preliminary actions, and producing the document.

Outline 19.1 Team Writing Process

Requirements
- Determine what has to be done to solve the problem.
- Ascertain what kind of document will be needed.
- Specify document themes and organization.

Preliminary actions
- Designate a leader, and provide adequate authority and responsibility.
- Determine what specific resources (skills, data, and material) will be needed.
- Assign teams and/or individuals to provide these resources.

Document production
- Define specific writing responsibilities and schedules.
- Task team members to produce draft inputs for their area(s).
- Edit and revise inputs as needed for the overall effectiveness of the document.
- Conduct a final review of all aspects of the document, and deliver as required.

Here is a brief discussion of this three-stage process.

Requirements

Normally, you would begin the team writing process with two basic assumptions. First, a need or opportunity exists that requires team action; and second, the output of the team action will be some type of solution laid out in a formal document. The first phase of this process, like all technical writing, involves defining the purpose of the document, whether that is fulfilling a course requirement or winning a large business contract. In other words, you need first to identify and understand the problem you are trying to solve before you jump into the process of solving it.

Next, you have to determine what kind of document is necessary. If you are trying to sell your company's goods and services, or convince a professor to approve your team's design topic, a proposal is in order. If you belong to a team of experts charged with looking into which alternative is best to solve an existing requirement in the company, a recommendation report is the right choice. If you are developing knowledge about a new, evolving technology that your company is interested in pursuing, a state-of-the-art research report might be just the thing.

In any case, once the type of document has been selected, the next step is to decide on the controlling ideas (themes) to be emphasized and the sections (organization) of the document. Normally, the style will be dictated by a style sheet, which often is provided either formally by a specified style guide or informally by the last successful project of this type that you can get your hands on. Copying the style from a previous document sometimes works, especially when a style guide is not available; but be careful not to clone the document itself. That rarely works well and often brings with it a host of legal and ethical issues, not to mention the possibility that you might be copying a lousy document. Finally, be sure to focus on themes—those recurring positive images of your solution, company, or team that need to be highlighted.

A Word about Themes

Themes, which are controlling ideas that recur throughout a document, play a powerful role in business writing, especially proposals. Proposals, as discussed in Chapter 6, are persuasive documents designed to sell goods and services, and as such, they totally succeed or fail depending on how well they convince the reader. Other technical documents, while not geared specifically to convincing someone of something, still have important persuasive aspects. For example, a research report is designed primarily to provide information. However, depending on how well it is written and how effectively it appeals to the reader, the report can also enhance or degrade the image or reputation of the individual or company that produces it.

Highlighting themes in technical writing is all about creating persuasive images. In other words, you want to predispose your reader to be persuaded by his or her emotions, or convinced by your credibility. The technical proposal discussed above may be built around a well-defined problem, an effective solution, a precise statement of work, and an impressive list of resources—but it can still benefit from positive themes such as quality work, solid performance, effective risk management, and creative problem solving.

Preliminary Actions

The first preliminary action in the team writing process is to select a leader and provide that person with the authority and responsibility needed to get the job done. A discussion of the philosophy of leadership is beyond the scope of this book, except that it is worth mentioning that there are two polar types of leadership: *democratic* and *autocratic*. Democratic team leadership assumes the members are capable of good judgment and in the long run will rise above petty behavior to do the right thing. This democratic method sounds good but usually does not work well in professional team writing environments where politics, pressure, and purse strings are often lurking in the background. This is not to imply that a Machiavellian approach is called for; but at a minimum, team writing leaders, once appointed or otherwise designated, must have the clear authority to make tough, often unpopular decisions.

Other preliminary actions involve determining what resources will be needed. This includes human resources (the creativity and skills necessary to pull off the project) and material resources (the facilities, data, and equipment needed to properly research, write, and produce the report). Finally, the last preliminary action is to assign teams or individuals to specific areas of responsibility. In a student group, you might be in charge of developing and writing the technical approach to solving the problem, while another student might be responsible for researching and locating the required resources.

Document Production

The last major part of the team writing process is the actual writing. This is the point where the real challenge often begins, because the team writing process is neither simple nor linear. In fact, many who should know better view the task of writing a report as consisting of three easy steps: getting information, organizing notes, and assembling the paper—as if the process were analogous to populating a circuit board and soldering the components in place. If only it were that simple! Team writing, in particular, can be a complex, demanding process complicated by almost anything going wrong—from the unexpected illness of a member, to petty disputes over who gets to do what.

The first step in getting any team to write well is to define clearly what everyone has to do. Each member must be aware of his or her responsibilities. Each member should know exactly what he or she is expected to do and when it has to be done. Additionally, each person should understand how he or she fits in with the total effort. No one in a writing team works in isolation. Everything is coordinated and interconnected. Failure on the part of one person or one subteam ultimately impacts everyone else and the success of the project.

In professional proposal writing, if the cost group fails to produce the cost volume on time, it really does not matter how good the proposed solution is or how innovative the management philosophy might be. The proposal will fail, and everyone will lose. Also, as a student, it is not hard to imagine the anxiety of approaching the last day of the last class before graduation, and having someone on your design team fail to produce a critical part of the required design report.

The bottom line: Members of writing teams must work hard at working together to get the job done. The team leader's role is to determine, often through consensus in a student group, the best way to organize and develop the required elements of the report. The team's role is to produce those elements. The leader's role in this regard is to identify common ground; create a structure that encourages cooperation and collaboration; clearly identify goals, themes, and expectations; and ensure quality, on-time performance.

The leader is also responsible for editing and revising team inputs. She or he must ensure that the assembled inputs respond fully and adequately to the original tasking, and include the required themes. In addition, the leader must ensure that the document is coherent, reads as a single work, has the right tone, and is consistent in its style and structure. Finally, the leader must provide configuration control of the document. The leader, and only the leader, should make or approve changes at this point. (You can understand now why many leaders lose their sense of humor by the end of this process.)

The last step involves the final review of all aspects of the document and delivery of the final product. The final review is sometimes affectionately known as an *idiot check* or *gut check*. The goal is to get one or more trusted, knowledgeable people (often called a *red team*) outside of the writing team to look over the document for dumb mistakes. These kinds of errors are easy to make, especially if one is close to the writing process—which everyone on the writing team is. Sometimes we all see what we want to see, with glaring errors going totally undetected; and sometimes we overlook illogical or inconsistent statements that are right in front of our eyes. That is why, in a professional setting where the stakes are high and the environment is unforgiving, writing teams often employ the services of seasoned copy editors to help out in this regard.

As an example of one of the most complex and demanding team writing activities in the professional business environment, we will focus on formal proposals involving the U.S. government (also see Chapter 6) and will associate this highly structured process with the major elements of the model provided in Outline 19.1 and schematized in Figure 19.1. A typical proposal team organizational chart is provided in Figure 19.2.

Example of Professional Team Writing

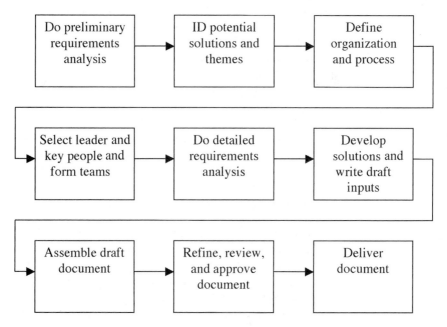

Figure 19.1
Professional team writing process.

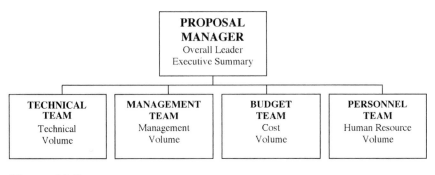

Figure 19.2
Typical proposal team.

Requirements

This team writing process usually begins when someone called a *Capture Manager* receives advance information about an upcoming project from either a Request for Information (RFI), in which the government basically asks private industry for ideas on solving a problem; or a draft Request for Proposal (RFP), in which the government gives a "heads up" to industry that it will be inviting industry to submit a proposal to solve some problem. The capture manager then analyzes the information (including whether the necessary funds are available), determines whether the project represents a good opportunity for his or her company, and, if so, defines the problem and roughs out how the company might respond with a proposed solution. The capture manager also defines the major themes for the proposal. These themes are recurring images that will be used deliberately throughout the proposal to highlight the strengths of the company. Specific themes in this case might include solid past performance, great technical talent, and superb facilities.

The capture manager effectively puts the team writing process into motion within his or her company. He or she prepares the company to receive the actual RFP when it is released and brings together both the human and the physical resources necessary to write the proposal. Often the capture manager will then become the *proposal manager*, the designated leader of the proposal effort. The proposal manager will be "on the hook" for the conduct and success of the entire project, and will serve as the document leader, the document coordinator, the editor of last resort, and the person with overall responsibility for the success of the project. When the proposal manager takes over, the role of the capture manager is finished.

Preliminary Actions

The proposal manager now assumes the primary leadership of the project. He or she determines how the project will proceed, what exactly will be done, what the schedule will be, and who exactly will do it. He or she will determine the need for various functional or technical teams, may rough out expected approaches to be taken by each team, and then will specifically task these teams. For example, the proposal manager would task the technical team to produce a refined, competitive technical solution to the problem and to draft the technical volume for the proposal. In a similar manner, he or she might task the management team to develop the organizational structure and management approach to support the technical solution, and to draft the management volume of the proposal. The same would be true for the accountants and financial managers producing the cost volume, etc.

The proposal manager then selects the various functional people who will serve in whatever specialty teams are needed, and may even appoint writers within the teams to draft their respective inputs to the proposal. The proposal manager might also recruit the services of technical writers and copy editors in a publications office, if one exists in the company. Either the proposal manager or the experts in the publications office would then define the style guide and procedures for actually producing the proposal.

Document Production

The proposal manager ensures that the RFP has been completely and thoroughly assessed and that every detail warranting a response has been identified and defined. (Failure to respond to even a single element could cost the company the contract.) A compliance matrix for each RFP requirement is built at this point. Ultimately, this matrix will be submitted to show evaluators where the RFP requirements are addressed in the proposal. Specific individuals or teams are then tasked to respond. The technical team, led by a technical team manager, develops the approach the company will propose to solve the problems defined by the RFP. Starting with the approach initially roughed out by the capture manager in the requirements determination phase, the technical team develops the proposed solution.

Of course, this solution cannot be developed in a vacuum. The technical approach has to be practical in terms of competitive cost and schedule, has to be consistent with the required management structure, and has to be doable in terms of the skills and resources available to the company. So a great deal of coordination must occur between the technical team and other teams. The proposal manager provides the primary conduit for this coordination, although individual team managers may also work closely with one another. During the document production phase, the proposal manager and individual team managers continue to provide adequate oversight and guidance and often have to make tough decisions on the spot. The emphasis must be on common sense and not having one team "spinning its wheels" while another team is making progress. Additionally, when individual teams submit their draft volumes, these drafts must be reviewed quickly and thoroughly, and then edited to make them consistent with the goals and themes of the overall proposal, as well as the content of other teams' drafts. Only the proposal manager makes or approves such changes at this point. The writing style and tone must be made consistent for all volumes. Needless to say, doing these things can be a very complex and demanding job, and proposal managers earn their pay!

Finally, the proposal manager assembles the entire document, including technical, management, cost, and human resource volumes, along with the compliance matrix—and in the process, adds an executive summary that he or she often writes. At this point, the entire project might receive a final review by a red team and approval by senior management. Additionally, the publications office may do final style and layout reviews. The document is then duplicated, and copies are delivered according to requirements in the RFP.

Student Team Writing

Student team writing differs from professional team writing. The experience base, skill levels, financial incentives, and authoritative structure are not present to the same degree in an academic setting as in a professional business environment. Additionally, student motivations are different. Obviously grades are on the line, but grade pressure varies from professor to professor and student to student. Some professors evaluate the entire group's performance, while others tend to hold each student more responsible for his or her contribution. Some students are strongly motivated to perform at high academic levels and would be upset with anything less than an A, while others may just want to get through the course and would be happy with any passing grade.

It is difficult to predict the dynamics and motives of any group of students involved in academic team writing. However, generally speaking, the process described in Outline 19.1 and schematized in Figure 19.3 is valid, and the principles described in this chapter do apply. Whether students are writing a proposal for a scientific study or producing a project report at the end of an engineering design sequence, writing team members must review the requirements of the project, take the necessary preliminary actions, and then produce the report.

Requirements

In an academic setting, there is no need for a capture manager or requirements expert. The teacher

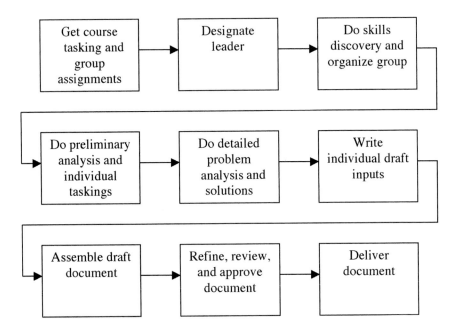

Figure 19.3
Student team writing process.

often defines the problem and provides the needed specifications. In some design sequences, however, students are assigned to groups and sent out to identify problems that they will then propose to solve. However, the teacher decides whether the problems they propose are worthy of being a course project and, specifically, what documents each group will need to produce to fulfill the course requirements.

Preliminary Actions

The preliminary actions phase is critical to the project's success. The first step for any student group is to designate a leader. In many cases, the selected leader is often the best student in the group—that is, the one who supposedly knows the most and who clearly has the most to lose. In other cases, the selected leader might even be the one person who did not attend the initial meeting and was not present to say no. However the selection is made, the leader should be a highly qualified member of the group who is willing and able to exercise authority and serve as the focal point for the project. If not, the group needs to find a new leader—it is just that simple! The leader then turns the group into a team by organizing and tasking the members. Unlike in the professional business environment, the type of leadership that works best with fellow students is usually more democratic than autocratic. So the leader needs to work first to establish consensus on the matters at hand. For example, if the teacher is leaving it up to the students to identify eligible problems, then group members need to agree on what kinds of problems should be considered and what resources would be required to solve these problems. As part of this process, the leader should also guide his or her fellow team members through an informal skills discovery process to determine what each individual brings "to the table" in terms of knowledge and experience.

The goal of this process is (1) to identify and define potential projects for the group to solve; (2) to determine to what extent each potential project is doable, and then select the best one; and (3) to accomplish the tasks associated with solving that problem by determining basically who will be responsible for doing what. In other words, the team member who is the best computer programmer would do any required coding, the most experienced "metal bender" in the group would build any required device or apparatus, and the most accomplished writer would take the lead on writing the report.

Document Production

Normally, to produce a report, each member submits a written input for his or her part of the project. The writer designated by the team takes these inputs and assembles a draft report. Assembling a report is much more involved than simply punching the pages and placing them into a three-ring binder. Assembly is a demanding, iterative process where the report writer interacts with the team leader, as well as each person submitting an input. The report writer organizes the material, fills in gaps, smooths out uneven prose, establishes a consistent tone, identifies missing parts, and makes other changes as necessary. The report writer and team leader obviously have to work very closely together during this phase of the process. It is not surprising that in many cases the team leader is the report writer.

Finally, once the project report has been assembled and a complete draft exists, each member reviews the *entire* document, not just his or her section. Often it is much easier for someone not closely involved with writing a particular section to spot errors or inconsistencies. Also, by this point in the process, the writer and/or team leader may be too close to each of the sections to notice problems. In fact, if possible, the group should enlist the aid of an outside reader who can review the entire report for factual or stylistic errors. The last step is for each member to review and approve the document for submission.

Conclusion

Team writing is a demanding process whereby several people, often with specialized skills and knowledge, are brought together to produce a single document. The collective efforts of a writing team (or teams) have the potential of providing great synergistic advantage, but the team can also break down and, in some cases, totally fall apart. Writing teams must have a clearly defined problem and clear management guidance from which to work. Before team writing begins, the specific requirements of the document must be clearly defined, the type of document must be determined, and the style and themes of the document must be specified.

Writing teams must also have a designated leader with overall responsibility for the project and with the authority and ability to make tough decisions. In professional team writing, this leader decides on the required skills and resources and assigns teams and individuals to accomplish the necessary tasks. In student team writing, the interaction is usually much less formal, and the team leader often works through consensus and negotiation. In any case, once the responsibilities for the project have been defined, the document can be produced. Specific writing tasks and schedules are laid out, and designated teams and members produce the required drafts. These drafts are then edited and revised as necessary, and the final review and production of the document occurs.

Index

A

'a' a lava, 273
Abstraction
  abstraction ladder, 2–4, 28
  carbon resistor, example of, 3
  concrete labels, 3
  precision, 5
  process of, 2–5
Abstracts and Summaries, 211–227
  abstracts, examples of, 212–214
  complete abstract, 212–214
  descriptive abstract, 211–212
  executive summaries, 214–227
  executive summary, example of, 216–227
  executive summary, writing strategy, 215–216
  informative abstract, 212–214
  limited abstract, 211–212
Academic standards, 250
Accuracy in technical writing, 8
Activity reports. See Progress report
Adobe Portable Document File (.pdf), 280–281
Air conditioner, 62–70
  compression, 64
  condensation, 64
  diagram of, 68
  evaporation, 65
  expansion, 65
Alteration of images, 21–23
Antenna, shortwave transmission, 205–209
  coverage, 207–208
  coverage map, 208
  dipole curtain, 207
  dipole curtain, diagram of, 207
  fan dipole, 206
  history, 206–207
  log-periodic, 206
  problem, 205–206
  theory, 206
Apparatus (laboratory report), 146–148
Appendices and attachments (proposal), 97–98
Application (cover) letter, 325–328

Approach statement (proposal), 88, 91, 100–101, 106
Asphalt road paving, 161
  electronic approaches to control, 161
  height and slope control, 160–161
  ultrasound, use in range finding, 162
Assessment, progress report, 114, 123
Attachments and appendices, 97–98
Audience, 5–10
  and prior experience, 4–5
  and purpose, 5–7
  report writing, 7–8

B

Background (research report), 195–196
Background statement (progress reports), 114
Background statement (proposal), 88, 90
Balloon help, 285
Bar and column charts, 266–267
Big Ears reconnaissance satellite, 136–137. See also Orbital transfer maneuvers
Bitmap (.bmp) file, 283
Black hole, definitions of, 33
Books, documenting, 253
BossRF Type 5000 transmitter, 141–143
Briefings, 297–314
  briefing charts typically included, 307–311
  briefing, definition of, 305
  briefings, technical, 305–313
  checklists for, 313–314
  complexity of, 311–312
  concluding chart, 310
  discussion chart, 308–309
  effectiveness factors, 298–301
  guidelines for, 305–307
  outline for, 306
  overview chart, 308
  presentations, 297–314

Briefings—*Cont.*
  presentations, definition of, 297
  speaking purposes, 303–305
  speaking situations, 301–303
  special effects, 312–313
  summary chart, 309–310
  title chart, 307

C

Capture manager, 346
Case errors, 238–239
Cause and effect, 31
Checklists
  briefing, 313–314
  cover letter, 334
  definition, 35–36
  documentation, 250
  electronic publishing, 294–295
  feasibility report, 144
  instructions, 189–190
  interview, 334–335
  laboratory report, 164–165
  mechanism description, 55
  presentations, 313–314
  process description, 77–78
  progress reports, 124
  project report, 164–165
  proposal, 108–109
  recommendation report, 144
  resume, 334
Classification (definition), 27–28
Comma splice errors, 230–231
Common file formats, 280–281
Common knowledge, 17, 48
Comparison and contrast, 31
Complete abstract, 212–214
Complexity of briefings, 311–312
Compliance matrix, 348
Compression (air conditioner), 64
Computer skills (resume), 321
Computer workstation design, 177–184
  chair adjustment, 178
  chair adjustment, diagram of, 180
  definition of, 177
  desk placement, 177
  desk placement, diagram of, 179
  lighting adjustment, 182–184
  work area layout, 181–182
Conceptual process, 59–60, 70–74
  compared to mechanisms in
    operation, 60
  outline for, 61–62
Concluding chart (briefings), 310
Conclusions, 47, 66, 94–95, 117,
  135, 152, 176, 198, 310
  instructions, 176
  laboratory reports, 152
  mechanism description, 47
  presentations and briefings, 310
  process description, 66
  progress reports, 117
  proposal, 94–95
  recommendation reports, 135
  research reports, 198

Condensation (air conditioner), 64
Conference papers, 254
Copyright, 249–250, 293
  and documentation, 249–250
  and electronic publishing, 293
  and Fair Use, 250
Cost statement, 93–94
Cost volume (proposal), 85
Cover letters, 96–97, 325–328
  checklist for (resume), 334
  example of (resume), 329
Credibility (documentation), 250

D

Definition, extensions, 30–33
  cause and effect, 31
  classification, 31
  comparison and contrast, 31
  etymology, 32
  exemplification, 32
  further definition, 31
  process, 32
Definition, technical, 25–38
  and audience, 29–30
  checklist for, 35–36
  classification for, 27–28
  classification using abstraction
    ladder, 27–29
  differentiation for, 29
  exercises, 36–37
  extensions for, 30–33
  model for, 25–29
  required imprecision, 33–34
  specifications and standards,
    34–35
Descriptive abstract, 211–212
  example of, 212
Diagrams, 263–264
Differentiation (definition), 29
Digital Millennium Copyright Act,
  294–295
Digital Versatile Disc (DVD), 284
Discussion chart, 308–309
Dissertations and theses, 255
Document production, 343–345
Documentation, 247–260
  academic standards, 250
  of books, 253
  checklist for, 259
  of conference papers, 254
  credibility, 250
  definition of, 248–249
  of dissertations and
    theses, 255
  of encyclopedias, 254
  of FTP sites, 258
  of images, 262
  of interviews, 258–259
  of journals, 254
  of lectures, 259
  legal requirements, 249–250
  of local computer storage, 258
  media (electronic), 253–259
  media (other), 258–259

media (print), 253–255
of newspapers, 254
of nonjournal entries, 255
notational, 250–251
of online forums, 258
parenthetical, 251
reasons for, 249–250
styles of, 249
of technical reports, 255
types and approaches, 250–251
of visuals, 262
of websites, 257–258

E

Economy of style, 244
Education (resume), 320
Effectiveness factors (briefings),
    298–301
Electronic publishing, 279–296
Adobe Portable Document File
    (.pdf), 280–281
Bitmap (.bmp), 283
checklist for, 294–295
common file formats, 280–281
definition of, 280
Encapsulated PostScript
    (.eps), 282
graphics file formats, 282–283
Graphics Interchange Format
    (.gif), 282–283
hyperlinked documents, 283–292
hypertext, converting to,
    285–287
Hypertext Markup Language
    (.htm/.html), 281
hypertext, organizing, 288–292
hypertext, production of, 285
image formats, 282–283
Joint Photographic Experts Group
    (.jpg/.jpeg), 282
Microsoft Word (.doc), 281
Picture (.pct/.pict), 283
plain text (.txt), 281
Portable Network Graphics
    (.png), 283
publishing and copyright, 293
Rich Text Format (.rtf), 281
Tag Image Format (.tif/.tiff), 282
TeX (.tex), 281–282
web browsers, 287–288
Encapsulated PostScript (.eps), 282
Encyclopedias, documenting, 254
Ergonomic design of computer
    workstations, 177–184
chair adjustment, 178
chair adjustment, diagram of, 180
definition of, 177
desk placement, 177
desk placement, diagram of, 179
lighting adjustment, 182–184
work area layout, 181–182
Ethical considerations, 11–23
ethics and technical writing, 12–20
exercises, 22–23

image alteration, 20–23, 275–276
importance of, 11
model for, 16
plagiarism, 17–20
Ethics
backdrop for technical writing, 11
definition of, 12
traditional constructs for, 13–14
Etymology (definition), 32
Evaporation (air conditioner), 65
Example topics
16XL1000000 Megatube, 29–33,
    54–55, 75–77, 105–108,
    122–123, 141–143,
    153–160, 185–188,
    205–209. *See also*
    Megatube 16XL1000000
air conditioner, 67–70
carbon resistor, 49–53
computer analysis capability,
    99–104, 118–121
ergonomic computer workstation
    setup, 177–184
Kilauea lava flow, 21–23
orbital transfer maneuvers,
    136–139
quantum-computing processor,
    199–204
road-paving operations, 160–164
selection sort algorithm, 71–74
shortwave antenna design
    considerations, 205–209
Examples
definition, 27
feasibility report, 141–143
instructions (expert), 185–188
instructions (layperson), 169–184
laboratory report (commercial),
    153–160
mechanism description
    (attributes), 49–53
mechanism description
    (functional), 54–55
process description (conceptual
    process), 71–73
process description (mechanism
    in operation), 67–70,
    75–77
progress report (commercial),
    118–121
progress report (student), 122–123
project report (student), 160–164
proposal (commercial), 99–104
proposal (student), 105–108
recommendation report, 136–140
research report, 199–204
state-of-the-art report, 199–204
team writing (professional),
    345–349
team writing (student), 349–353
Executive summary, 214–227
definition of, 214–215
example of, 216–227
Executive summary (proposal), 85
Exemplification (definition), 32

Exercises, 21–23, 35–37, 56–58,
    78–82
exercise, definition, 36–37
exercise, ethics, 21–23
exercise, mechanism description,
    56–58
exercise, process description, 78–82
Expansion (air conditioner), 65
Experience (resume), 322
Extensible Markup Language
    (XML), 284

F

Facilities and equipment (proposal),
    92–93
Fair Use (copyright), 250
Feasibility reports, 125–144
    candidate solutions
        identification, 128
    checklist for, 144
    conclusions and
        recommendations, 129
    criteria development, 128
    data collection, 128–129
    definition of, 125
    differences from recommendation
        reports, 125–126
    example of, 141–143
    Hohmann transfer, 137–138
    one-tangent burn, 137–138
    outline for, 126
    problem definition, 127, 130
    purpose statement, 129–130
    scope statement, 131–132
    threaded example, 141–143
FinkelBOAT, 56–58
    image of, 57
Formal proposals, 84–87
FTP site, documenting, 258
Functional mechanism description,
    53–55
Further definition (definition), 31
Fused sentence errors, 231–232

G

Government proposals, 86–87, 345
Grammar and style, 229–246
    case, 238–239
    comma splices, 230–231
    common errors, 230–243
    definition of, 229
    economy of style, 244
    fused sentences, 231–232
    homonyms, 240–241
    importance of, 229–230
    misplaced modifiers, 233–234
    noun clauses, 242–243
    numbers, 241–242
    passive voice, 234–235
    precision of style, 245
    pronoun agreement, 237–238
    pronoun reference, 238
    sentence fragments, 232–233

spelling, 239–240
    stylistic considerations, 243–246
    verb agreement, 236
Graphics Interchange Format (.gif),
    282–283
Graphs, 264
Group vs. team writing, 337–338
Group writing. *See* Team writing
Guidelines for briefings, 305–307

H

Help files, 285
Hohmann transfer, 137–138
Homonyms, 240–241
Hyperlinked documents, 283–292
Hypertext
    converting to, 285–287
    organizing, 288–292
    production of, 285
Hypertext Markup Language
    (.htm/.html), 281

I

Image alteration, 21–23
Image compositing, 21–23
    Kilauea volcano, 22–23
Image formats (electronic
    publishing), 282–283
Images, documenting, 262
Informal proposals, 88–89
Informative abstract, 212–214
Instructions, 167–190
    checklist for, 189–190
    definition of, 167
    outline for, 168
Interviewing for job, 331–333
    checklist for, 334–335
Interviews, 315–336
    documentation of, 258–259
Introduction, technical writing, 1–10

J

Job application, 315, 328
Job objective (resume), 318
Joint Photographic Experts Group
    (.jpg/.jpeg), 282
Journals (documentation), 254

K

Kilauea volcano, 21–23
Kinetic energy equation, 135, 139

L

Laboratory reports, 145–165
    apparatus and procedure, 151
    assessment, 152
    background for, 150
    checklist for, 164–165
    definition, 146–147
    differences from project report, 146

example of, 153–160
findings, 151–153
old horror movies, 145
outline for, 147
project reports, compared with,
146–147
test and evaluation, 150–151
threaded example, 149–153
Ladder of abstraction, 2–4, 28
Lava, volcanic (Kilauea)
'a' a lava, 273–274
pahoehoe lava, 273–274
Layout and presentation (proposal), 95
Lecture, documentation of, 259
Legal requirements for
documentation, 249–250
Limited abstract, 211–212
Links, top level, 288
Loading time reduction, 287–288
Local computer storage,
documentation of, 258
Love, 6–7
scientific definition of, 7
Love note, 6
creative version, 6
technical version, 6
List of references, 251–253

M

Macromedia Dreamweaver, 291
Management volume (proposal), 85
Manuals, 167–190
differentiated from instructions,
188–190
Markup language, 281
Mechanism description, 39–58
16XL1000000 diagram, 76
16XL1000000 schematic, 75
16XL1000000 specification sheet,
54–55
33-kilohm, 1-watt carbon resistor,
49–53
checklist for, 55
definition of, 39
exercise, 56–58
FinkelBOAT, diagram of, 57
FinkelBOAT exercise, 56–58
functional, 53–55
outline for, 40
resistor example, 41–53
threaded example, 53–55
visuals, 48–49
Mechanism in operation, 61, 67–70.
See also Process description
Media
electronic, 253–259
other, 258–259
print, 253–255
Megatube 16XL1000000, 29–33,
54–55, 75–77, 105–108,
122–123, 141–143, 153–160,
185–188, 205–209. *See also*
Neutralization (16XL1000000)
collection, 77

control, 77
diagram of, 76
emission, 76
neutralization of, 185–188
operation of, 75–77
schematic of, 75
specifications, 54–55
use in audio applications,
153–160
Microsoft Word (.doc), 281
Milestone reports, 111–112
Misplaced modifiers, 233–234

N

Neutralization (16XL1000000),
185–188
adjustment, 188
circuit diagram, 186
definition of, 185
high-voltage plate tap, 187
input inductor, 186
neutralization capacitor, 189
Newspapers (documentation), 254
Nonjournal entries, 255
Notational documentation, 250–251
Numerical data, reliance on, 8

O

One tangent burn, 138–139
Online forum (documenting), 258
Operation of a mechanism, 59–70.
See also Process description
Orbital transfer maneuvers, 136–140
accuracy, 137–138
diagram of, 138
geosynchronous earth orbit,
136–137
Hohmann transfer, 137
kinetic energy, 139
near earth orbit, 136
one tangent burn, 137
required fuel, 139–140
time of flight, 138–139
Outlines, 40, 61–62, 88–89, 114,
126–127, 306, 317–318, 340
briefing, 306
feasibility reports, 126
informal proposal, 88–89
mechanism description, 40
process description, conceptual
process, 61–62
process description, mechanism
in operation, 61
progress reports, 114
recommendation reports, 127
resume, 317–318
team writing, 340
Overview chart (briefing), 308

P

Pahoehoe lava, 273–274
Parenthetical documentation, 251–253

Parenthetical references, 251–252
Personal information (resume), 323
Personnel resources (proposal), 92
Photographs, 273–277
    alteration of, 275–276
    ethics of altering, 20–23
Pictographic charts, 269, 272
Picture (.pct/.pict), 283
Pie charts, 268
Plagiarism, 17–20
    common knowledge, 17
    definition of, 17
    detection risk, 18
    ethical imperative, 18
    and the Internet, 17–18
Plain text (.txt), 281
Portable Network Graphics
    (png), 283
Powerpoint, Microsoft, 305–313
Precision of style, 245
Presentations and briefings,
    297–314
    checklist, general, 313
    checklist, technical briefings,
        313–314
    complexity, controlling, 311–312
    concluding chart, 310
    definition of, 297
    demonstrative, 304
    discussion chart, 309
    effective delivery, 300–301
    extemporaneous, 302–303
    guidelines for, 305–310
    impromptu, 301–302
    informative, 304
    manuscript, 303
    overview chart, 308
    persuasive, 304
    speaking purposes, 305
    speaking situations, 301–303
    special effects, 312–313
    substantive ideas, 298
    summary chart, 309–310
    terminology and concepts, 299
    title chart, 307
Problem definition (feasibility and
    recommendation reports),
    127, 130
Process (definition), 32
Process description, 59–82
    conceptual process, 70–74
    conceptual process, outline for,
        61–62
    definition of, 59–60
    exercise, 78–82
    mechanism in operation, 67–70
    mechanism in operation, outline
        for, 61
    outlines for, 61–62
    QuadFINKEL, diagram of, 80
    QuadFINKEL exercise, 78–82
    selection sort, 71–73
    threaded example, 75–77
    visuals, 66

Process limitations (visuals),
    276–277
Progress report, 111–124
    background statement, 114
    checklist for, 124
    conclusion, 117
    definition of, 112
    example, commercial, 118–121
    example, student, 122–124
    formats for, 113
    outline for, 114
    project time line chart, 121
    purpose statement, 113
    scope statement, 115
    status statement, 116–117
    threaded example, 122–124
Project report, 145–165. *See also*
    Laboratory reports
    apparatus and procedure,
        162–163
    assessment, 164
    background, 161–162
    checklist for, 164–165
    definition, 146–147
    differences from laboratory
        reports, 146
    example of, 160–164
    findings, 163–164
    outline for, 148
    test and evaluation, 162–163
    threaded example, 149–153
Pronoun agreement, 237–238
Pronoun reference, 238
Proposal, checklist, 108–109
Proposal, example (commercial),
    99–104
Proposal, example (student),
    105–109
Proposal manager, 347
Proposal, threaded example,
    105–109
Proposals, 83–108
    appendices and attachments,
        97–98
    approach statement, 91
    background statement, 90
    conclusion, 94–95
    cost volume, 85
    costs statement, 93–94
    cover letters, 96–97
    definition of, 83
    executive summary, 85
    facilities and equipment, 92–93
    formal proposals, 84–86
    government proposals, 86–87
    informal proposal, 87–104
    layout and presentation, 95
    management volume, 85
    outline for, 88–89
    personnel (human) resources, 92
    purpose statement, 89
    request for proposal (RFP), 84–86
    resources volume, 85
    result statement, 91–92

scope statement, 90–91
solicited vs. unsolicited, 87
statement of work, 92
technical volume, 85
title pages, 96–97
Publishing and copyright, 293
Purpose statement (progress
report), 113

Q

QuadFINKEL, 78–82, 99–104,
118–121
diagram of, 80
difficulty in modeling, 90
Quantum Chips QCPU, 193–204
background, 199–200
diagram of, 203
genesis, 201–202
theory, 200–201
Quantum computing, 199–204

R

Recommendation report, 125–144
candidate solutions
identification, 128
checklist for, 144
conclusions and
recommendations, 129
criteria development, 128
data collection, 128–129
definition of, 125
differences from feasibility
reports, 125–126
example of, 136–140
Hohmann orbital transfer, 137–138
one-tangent burn orbital transfer,
137–138
outline for, 127
problem definition, 127, 130
purpose statement, 129–130
scope statement, 131–132
Red team, 344
Request for proposal (RFP), 84–86
Required imprecision, 33–34
Requirements phase (team writing),
340–341
Research reports, 191–210. See
also State-of-the-art reports
background, 195
checklist for, 210
conclusion and summary, 198
definition of, 191–192
discussion section, 196–197
outline for, 192–193
problem statement, 194
purpose statement, 193
references and appendix,
198–199
state-of-the-art report example,
199–204
threaded example, research
report, 205–209

Resistor (mechanism description),
41–53
Resources volume (proposal), 85
Resumes and cover letters, 315–336
cover letter checklist, 334
cover letter example, 329
cover letters, 325–328
interview checklist, 334–335
interviewing, 331–333
resume checklist, 334
resume, definition of, 316–317
resume example, 330
resume outline, 317–318
resume writing, 317–323
resumes and the Internet, 328
ten tips for good resumes,
323–325
Rich Text Format (.rtf), 281
Road paving operations, 160–164
asphalt road paving, 161
control of height and slope,
160–161
electronic approaches to
control, 161
ultrasound, use in range
finding, 162

S

Satellite, Big Ears reconnaissance,
136–137. See also Orbital
transfer maneuvers
Schematics, 269
Selection sort, 71–74
example of, 74
process description, 70–74
schematic flow chart, 72
Sentence fragments, 232–233
Shortwave antenna design,
205–209
coverage, 207–208
coverage map, 208
dipole curtain, 207
dipole curtain, diagram of, 207
fan dipole, 206
history, 206–207
log-periodic, 206
problem, 205–206
theory, 206
Solicited vs. unsolicited
proposals, 87
Sort algorithm, process description
of, 70–74
Source citation, 251–252
Source line for visuals, 262
Speaking purposes, 303–305
Speaking situations, 301–303
Special effects (briefing), 312–313
Specification and standards, 34–35
Spelling, 239–240
Stall, definitions of, 25–27
Standard Generalized Markup
Language (SGML), 284
Statement of work (proposal), 92

State-of-the-art reports, 191–210
  background, 195
  checklist for, 210
  conclusion and summary, 198
  definition of, 191–192
  discussion section, 196–197
  example of, 199–204
  outline for, 192–193
  problem statement, 194
  purpose statement, 193
  references and appendix, 198–199
Status reports, 111–112
Status statement (progress report),
  116–117
Strengths statement (resume),
  319–320
Style, 229–246
Styles of documentation, 249
Stylistic considerations, 243–246
Summaries (executive), 211–228
Summary chart (briefing), 309–310
Summary table (recommendation
  report), 140

T

Tab Image Format (.tif/.tiff), 282
Tables, 270–271
Team writing, 337–353
  compared to *group writing,* 337
  definition of, 337–338
  document production, 343–345
  example of (professional), 345–349
  example of (student), 349–353
  outline for, 340
  preliminary actions, 342–343
  process of, 339–345
  requirements phase, 340–341
  student vs. professional, 338–339
  themes, 341–342
Technical definition, 25–38
Technical reports (documentation),
  255
Technical volume (proposal), 85
Technical writing
  audience, 3–7
  definition of, 1, 7–8
  measure of merit for, 5
  precision, 3
  purpose, 5–7
Technical writing and abstraction, 2–3
Ten tips for good resumes, 323–325
TeX (.tex), 281–282
Threaded examples, 53–55, 75–77,
    105–109, 122–124, 141–143,
    153–160, 185–188, 205–210
  feasibility report, 141–143
  instructions, 185–188
  laboratory report, 153–160
  mechanism description, 53–55

  process description, 75–77
  progress reports, 122–124
  proposal, 105–109
  state-of-the-art report, 205–210
Three dimensional (3-D) charts, 268
Time, creative and technical
  definitions, 1–2
Title chart (briefing), 307
Title pages, 96–97
Top level links, 288
Traditional document organization,
  290
Transmitter, BossRF Type 5000,
  141–143

U

Universal resource locator
  (URL), 287
Unsolicited proposals, 87
U.S. government requirements
  (proposals), 86

V

Vacuum tube theory, 154–155
Velocity, 139
Verb agreement, 236
Village Thumpers, 149
Virtual Reality Markup Language
  (VRML), 284
Visuals, 261–278
  adding interest to, 268–269
  bar and column charts, 266–267
  definition of, 261
  diagrams, 263–264
  graphs, 264–266
  guidelines for design, 262–263
  guidelines for use, 261–262
  image alteration, 20–23, 275–276
  photographs, 273–277
  pictographic charts, 269, 272
  pie charts, 268
  process limitations, 276–277
  schematics, 269
  tables, 270–271
  three dimensional (3-D)
    charts, 268
  types of, 263
Visuals, reliance on, 7–8
Volcanic lava, 273–274
Volcano, Kilauea, 21–23, 273–274

W

Web browsers, 287–288
  control of appearance, 288
  file loading considerations, 287
  universal resource locator, 287